The Edge of Infinity
Supermassive Black Holes in the Universe

In the past they were recognized as the most destructive force in nature. Now, following a cascade of astonishing discoveries, supermassive black holes have undergone a dramatic shift in paradigm. Astronomers are finding out that these objects may have been critical to the formation of structure in the early universe, spawning bursts of star formation, planets, and even life itself. They may have contributed as much as half of all the radiation produced after the Big Bang, and as many as 200 million of them may now be lurking through the vast expanses of the observable cosmos. In this elegant nontechnical account, Melia conveys the excitement generated by the quest to expose what these giant distortions in the fabric of space and time have to say about our origin and ultimate destiny. This fascinating and timely book is suitable for the general reader wishing to find answers to some of the intriguing questions now being asked about black holes.

FULVIO MELIA is Professor of Astronomy and Associate Head of Physics at the University of Arizona. He has held numerous visiting appointments at universities in Europe and Australia, including posts in Paris, Heidelberg, Padua, and Melbourne. Melia has won many national and international distinguished awards for scholarship and teaching. His research focuses on the characteristics and behavior of black holes and other compact stars, and he has published over 180 papers in the primary literature.

The Edge of Infinity

Supermassive Black Holes in the Universe

FULVIO MELIA

The University of Arizona, Tucson

CAMBRIDGE
UNIVERSITY PRESS

PUBLISHED BY THE PRESS SYNDICATE OF THE UNIVERSITY OF CAMBRIDGE
The Pitt Building, Trumpington Street, Cambridge, United Kingdom

CAMBRIDGE UNIVERSITY PRESS
The Edinburgh Building, Cambridge CB2 2RU, UK
40 West 20th Street, New York, NY 10011–4211, USA
477 Williamstown Road, Port Melbourne, VIC 3207, Australia
Ruiz de Alarcón 13, 28014 Madrid, Spain
Dock House, The Waterfront, Cape Town 8001, South Africa

http://www.cambridge.org

© Cambridge University Press 2003

First published 2003

Printed in the United Kingdom at the University Press, Cambridge

Typeface Trump Mediaeval 9.5/15 pt.   *System* LATEX 2_ε [TB]

A catalog record for this book is available from the British Library

Library of Congress Cataloging in Publication data
Supermassive Black Holes / Fulvio Melia.
   p.   cm.
Includes bibliographical references and index.
ISBN 0 521 81405 7
1. Black holes (Astronomy)   I. Title.
QB843.B55M455   2003
523.8′875 – dc21   2003044036

ISBN 0 521 81405 7 hardback

Contents

Preface

If you were to imagine a description of nature whose constituents are so bizarre that even its originator refuses to allow for their actual manifestation, you would not have to go past the theory of general relativity. Created almost a century ago, it was perhaps the most anticipatory advancement in the history of physics. Its development was so visionary that none of the four significant tests applied to it since – two of which were adjudged to be of Nobel quality – have truly exposed the core of this remarkable theory, where the most abstruse distortions to the fabric of space and time are imprinted.

Einstein suspended his belief at the thought of a universe that would permit singularities to form, in which matter collapses inexorably to a point and becomes forever entombed. Yet this was the boldest consequence of his new description of gravity.

Remarkably, the idea that a gravitational field ought to bend the path of light so severely that the heaviest stars should then be dark was actually forged much earlier, in the context of Newtonian mechanics, toward the end of the eighteenth century. The Reverend John Michell argued in a paper published by the Philosophical Transactions of the Royal Society that, if a star was sufficiently massive, its escape velocity would have a magnitude exceeding even the speed of light, which, being comprised of particles, would then slow down and fall back to the surface. These stars would therefore be unobservable and he coined the term "dark star" to vividly portray this peculiar property. (Incidentally, this discussion appears to have been the first mention of the possible existence of dark matter in the universe.)

It is indeed a measure of how extraordinary such objects are that it took almost 200 years before some evidence for the existence of "black holes," as they are now known, started trickling in. These days, astronomers are riding a cascade of astonishing discoveries, many of them with space-based facilities such as the Hubble Space Telescope and the Chandra X-ray Observatory, and are finding themselves on the other side of a dramatic shift in paradigm.

Twenty years ago, the idea of giant black holes the size of our solar system seemed more like fodder for science fiction than something relevant to the real world. Widely recognized as the most destructive force in the universe, supermassive black holes could not easily fit into the highly ordered structure that astronomers saw in galaxies and their clustering. But now we are finding out that these objects may have contributed as much as half of all the radiation pervading the intergalactic medium, and that as many as 200 million of them may be lurking in the vast expanses of the known universe.

Exactly how galaxies were created continues to puzzle astrophysicists, who grapple with the question of why the primordial gas collapsed to form the aggregates of matter we see today. It is starting to look more and more as though supermassive black holes were critical in this process. Their overwhelming gravity may have triggered condensations that eventually led to the majestic cartwheels of spiral galaxies such as the Milky Way. They may have spawned bursts of star formation, planets, and, yes, even life itself. So supermassive black holes may have been here at the beginning and, because they possess a one-way membrane that draws matter in, but lets nothing back out into the universe as we know it, they will all be here toward the end.

The exhilarating black-hole discoveries produced by the astronomical community invigorate the public's imagination. This is the element that has motivated the writing of this book. For her patience and support throughout the course of this project, I am indebted to Jacqueline Garget, who early on saw the need for such a treatise. And for generously supporting my research in this area for over a decade and a half, I gratefully acknowledge the National

Science Foundation, the National Aeronautics and Space Administration, and the Alfred P. Sloan Foundation.

Finally, I owe a debt of gratitude to the pillars of my life – Patricia, Marcus, Eliana, and Adrian – and to my parents, whose guidance has been priceless.

1 The most powerful objects in the universe

History's time line swept through 1963 with a breathtaking pace. The community of nations was about to welcome the birth of its newest member, Kenya, which that year attained independence from Great Britain. The Vietnamese military, meanwhile, was in the process of overthrowing the regime of Ngo Dinh Diem, deepening the US involvement in Southeast Asia and setting the stage for a decade of discordant relations among the superpowers. Ironically, this was also the year in which the first test ban agreement between the USA and the Soviet Union was ratified, concluding a nervous endeavor to ease growing nuclear tensions. For the individuals in society, the issue of women's rights resurfaced, promoted by Betty Friedan's just-released book *Feminine Mystique*. And while readers were being exposed to the idea of a modern woman discarding her traditional role, humanity as a whole was gaining some leverage over nature with the discovery of a vaccine against the measles. Many remember 1963 for the tragic assassination of President John F. Kennedy.

This tessellation of historical markers stirring the world in 1963 formed quite a backdrop for two minor events that would lead, over time, to the eventual uncloaking of the most powerful objects in the universe. At Mount Palomar Observatory, Maarten Schmidt was pondering over the nature of a starlike object with truly anomalous characteristics, while Roy Kerr, at the University of Texas, was making a breakthrough discovery of a solution to Albert Einstein's (1879–1955) general relativistic field equations. Kerr's work would eventually produce a description of space and time surrounding a spinning black hole, which is now thought to power very dense concentrations of matter like those responsible for producing the mystery on Schmidt's desk in 1963. Perhaps the most enigmatic objects in the cosmos, black holes enclose regions of space within which gravity is so strong that not even light can escape – hence their name. Light paths originating near them are bent by the strong gravitational field and wend their way inwards toward oblivion, creating a dark depression in an otherwise bright medium.

The astronomical puzzle on Schmidt's desk was the star recently associated with the 273rd entry in the third Cambridge catalog of radio sources, hence its designation as 3C 273. For centuries, such objects had gone unnoticed, appearing in the nighttime sky merely as faint points of light. The development and use of radio telescopes in the 1940s, however, led to the gradual realization that several regions of the cosmos are very bright emitters of centimeter-wavelength radiation. As 1963 approached, the British astronomer Cyril Hazard devised an ingenious method of pinpointing the exact location of such a source. Using lunar occultation, he suggested, it should be possible to note the precise instant that the radio signal stopped and then re-emerged when the moon passed in front of it. Astronomers could then determine with which, if any, of the known visible objects in the firmament the emitter of centimeter-wavelength radiation was associated.

Hazard arranged to make the measurements at Parkes Radio Telescope situated several hundred miles from the University of Sydney in the Australian outback. But the observation that he had proposed almost did not happen. He took the wrong train that night and missed the event entirely. Fortunately, the staff at the observatory, headed by the director John Bolton, proceeded with the plan anyway. It turned out to be a rather daring feat since the region to be observed was too close to the horizon, and the telescope could not tip over sufficiently to make the recording. Undaunted, the observatory staff cut down the intervening trees and removed the telescope's safety bolts, allowing the several-thousand-ton facility to swivel sufficiently to catch the occultation.

Hazard's experiment worked beautifully, and the radio source tracked by Parkes that night – 3C 273 – could be identified with a single starlike object in the constellation Virgo. It seemed like a rather docile object, but its appearance belied the fact that this quasi-stellar radio source (hence the name "quasar") is a prodigious emitter of radiation. The characteristics of its optical light – basically, the colors of its rainbow – had never been seen before. Schmidt eventually solved the puzzle by realizing that the pattern of colors before him was really that produced by hydrogen atoms, only with a wavelength shifted by about 16 percent from its value produced in the laboratory.[1] But was this simply an indication that the physics in space was unusual compared to what we see on Earth?

1.1 BEACONS AT THE EDGE OF REALITY

The answer, it turns out, really had to do with the fact that 3C 273 was moving away from us. Just as the pitch of a whistle depends on how fast the train is receding or approaching, the shift in wavelength of light is an indicator of the speed with which its source is moving. The greater is its redshift – as the increase in wavelength is called – the higher is its speed of recession. (Similarly, a "blueshift" would indicate that the source was approaching us.) To Schmidt and his colleagues, the shift in wavelength was quite remarkable, for it had been known since the time of Edwin Hubble (1889–1953), the great astronomer for whom the Space Telescope is named, that cosmological distances scale directly with speed. According to the redshift that Schmidt had identified, 3C 273 had to be much farther away than had been previously imagined.

To understand the origin of this interpretation, we must turn the clock back some 40 years, to a period when the idea of a "universe" had not yet been fully gestated. Prior to the 1920s, most of Hubble's colleagues believed that the Milky Way galaxy, the swirling collection of stars that fills the night sky in the southern hemisphere, was essentially the entire cosmos. Moving at 250 kilometers per second, it takes

---

[1] Maarten Schmidt reported his discovery in a one-page article published by *Nature* in 1963.

our Sun about 220 million years to complete one orbit about the center of this structure, which was therefore viewed as being of sufficient size to satisfy our deep-rooted yearning for the universe to be immense compared to human proportions. But looking into deep space from the chilly summit of Mount Wilson, in Southern California, Hubble realized that the Milky Way is instead only one of many galaxies fighting the darkness in an incomparably larger cosmos.

Toward the end of the 1920s, as the world inched closer to the precipice of a great depression, Hubble surprised the scientific community with yet another remarkable discovery: the galaxy-studded universe, he claimed, was actually *expanding*. Like dots on a swelling balloon, these bright markers were receding from each other with a speed that increased with distance. Even Albert Einstein (see Fig. 1.1), who had earlier postulated that the universe was static and eternal, was caught by surprise and felt compelled in 1929 to acknowledge and retract what he termed "the greatest blunder of my life." Hubble's discovery is now viewed as the first evidence for the Big Bang theory of creation, in which the known universe not only had an origin in time, but apparently began its inexorable expansion as a single point in space, within which all matter and energy were initially compressed. This revelation turned Hubble into a worldwide celebrity, becoming a favored guest in Hollywood during the 1930s and 1940s, where he befriended the likes of Charlie Chaplin, Helen Hayes, and William Randolph Hearst.

It was already known that the light from certain strange-looking nebulae (at the time, these had not yet been associated with galaxies, but were instead believed to be merely glowing balls of gas) was redder than it should be. And motion away from us seemed to be the most likely cause for this so-called redshift. Now, with meticulous care, Hubble was documenting the distance to these receding nebulae and found what has come to be known as Hubble's Law: that the farther away the galaxy is from Earth, the faster it appears to be moving.

For discovering the universe and inventing the field of cosmology, Hubble was rightfully the recipient of many awards, but the one

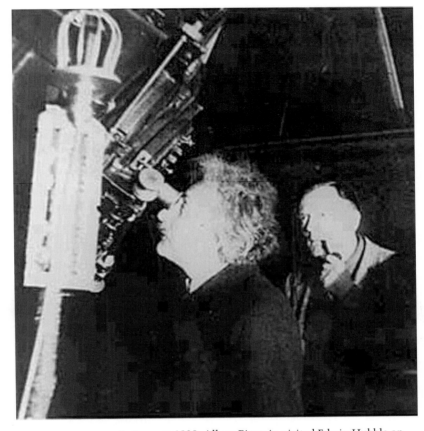

FIGURE 1.1 In January 1930, Albert Einstein visited Edwin Hubble on Mount Wilson, where the discovery of an expanding universe had been made. In this photograph, Hubble (in the background) looks on as Einstein peers through the Newtonian focus of the 100-inch telescope. (From the Edwin Hubble Papers. Reprinted with the kind permission of the Huntington Library, San Marino, California)

recognition that eluded him to his death was the Nobel Prize. This was not for lack of effort, however, for in the late 1940s he even hired an agent to publicize his meritorious work. Unfortunately, by the time the Nobel committee added the field of astronomy (in which he would have won the prize) to the eligible branches of physics, it was too late. He died in 1953. But his name lives on, gracing the sides of the Space

Telescope which is successfully carrying on the great work that he started on a cold mountain top in Southern California.

And so, three decades after Hubble's discovery, Schmidt had no trouble convincing his colleagues that the 16 percent redshift inferred from 3C 273's light implied a speed of recession of almost 30 000 miles per second and, therefore, a distance of three billion light-years from Earth.[2] Astronomers thus concluded that its starlike image must clearly be concealing its true nature; it had to be among the most powerful emitters of radiation in the universe in order for it to stand out so vividly over such a large cosmic expanse.

A recent image of this historic object was made with the Chandra X-ray telescope (see Fig. 1.2). Formerly known as the Advanced X-ray Astrophysics Facility, this state-of-the-art detector was launched in 1999 aboard the Space Shuttle, and was renamed the Chandra X-ray Observatory in honor of the late Indian-American Nobel laureate, Subrahmanyan Chandrasekhar. The word Chandra, which means "moon" or "luminous" in Sanskrit, is a very fitting name for this mission, recognizing Chandrasekhar's tireless pursuit of knowledge and understanding. Much of his work was devoted to developing a theory of black holes and other phenomena that the Chandra X-ray Observatory is now studying. He is widely regarded as one of the foremost astrophysicists of the twentieth century, winning the Nobel Prize in 1983 for his theoretical work on the physical processes that govern the structure and evolution of stars. With the ability to resolve features fifty times smaller than previous missions, Chandra is revolutionizing X-ray astronomy, and at more than 15 meters in length and weighing more than 50 tons, it is one of the largest objects ever placed in Earth orbit by the Space Shuttle.

The quasar story does not end here, however, for having crowned 3C 273 as one of the most powerful objects in the universe, astronomers were yet to uncover its most surprising characteristic. They

---

[2] A light-year is the distance light travels in one year. By comparison, it takes light a mere eight minutes to reach us from the Sun.

soon realized that variations in the total light output of 3C 273 and its brethren were occurring over a period of only 10 to 20 months, implying that the size of the region producing the optical light could not exceed a few light-years – basically the distance between the Sun and its nearest stellar neighbor. Imagine a crowd of people waiting to see a famous actor appearing somewhere in a plaza. Suddenly someone at one end of the enclosure makes the sighting and quickly spreads the news to the others. By the time everyone in the crowd has been alerted and has turned to face the celebrity, the number of seconds that will have passed depends on how quickly it takes the news to pass along from person to person. The bigger the crowd, the longer it takes for everyone to reorient their attention toward the actor. In 3C 273, the fastest conceivable signal travels at the speed of light, so this powerful source of radiation could not be bigger than the distance over which light will have traveled during the 10 to 20 months of observation. For this reason, astronomers infer that the most powerful objects in the universe are extremely small compared with even just the Sun's neighborhood, let alone a structure as big as the Milky Way, which stretches over a distance 100 000 times bigger.

Space-borne X-ray detectors such as Chandra have reinforced these conclusions by demonstrating that quasars are more luminous still in X-rays than they are in optical light. Their total X-ray output can vary over a period of only hours, corresponding to a source size smaller than Neptune's orbit. In fact, quasars are the most powerful emitters of X-rays yet discovered. Some of them are so bright that they can be seen at a distance of 12 billion light-years. Each quasar typically releases far more energy than an entire galaxy, yet the central engine that drives this powerful activity occupies a region smaller than our solar system!

Now, four decades after these intriguing objects were first identified on the basis of their optical light, astronomers are even discovering quasars whose radiative output is for the most part so feeble that they would not otherwise have been detected, except that these objects just happen to be the most powerful gamma-ray sources in the universe. Gamma rays are the most energetic photons – individual bundles of energy that together give substance to a beam of light – we have so far been able to detect with space-based instruments. They are very difficult to produce in numbers and very few cosmic objects can radiate at this energy with sufficient strength for us to be able to sense their presence. Recently, an international team of scientists using data from NASA's Compton Gamma-Ray Observatory, launched in 1991, has uncovered a quasar that blazes the heavens with this type of radiation from the far reaches of the universe, some 10 to 11 billion light-years away, though it barely flickers in the visible light range.[3] This remarkable feat was accomplished by combining images from successive scans of a patch of sky in the Ursa Major constellation, where the quasar is located.

This fantastic idea – that such a small volume could be screaming across the universe with the light of 100 billion Suns – has led physicists to conclude that quasars must be the radiative manifestation of supermassive black holes. In the next two chapters, we will learn about the exquisite detective work that has provided us with valuable information concerning the mass of a quasar and its incomparable power.[4]

It is natural to wonder whether these objects are "naked" – deep, dark pits of matter floating aimlessly across the primeval cosmic soup – or whether they are attached to more recognizable structures in the early universe. The answer to this question came when the strong non-stellar light from the central quasar was eliminated using mechanical and electronic means. In a few cases, a fuzzy haze was seen surrounding the bright beacon and, when this light was examined carefully, it turned out to have the colors and other characteristics of a normal giant galaxy.

In recent years, the task of source identification has been made easier using the Hubble Space Telescope (see, for example, Fig. 1.3). Resolving the mystery concerning the nature of quasars was in fact

---

[3] See Malizia *et al.* (2001) for a technical account of this very interesting discovery.
[4] See, for example, Salpeter (1964), Zel'dovich and Novikov (1967), and Lynden-Bell (1969).

FIGURE 1.3 This Hubble Space Telescope image reveals the faint host galaxy within which dwells the bright quasar known as QSO 1229 + 204. The details in this picture help solve a three-decade-old mystery about the true nature of quasars, the most distant and energetic objects in the universe. The quasar is seen to lie in the core of an ordinary-looking galaxy with two spiral arms of stars connected by a bar-like feature. This image shows one of a pair of relatively nearby quasars that were selected as early targets to test the resolution and dynamic range of the Hubble's then newly installed Wide Field and Planetary Camera, which contained special optics to correct for a flaw in Hubble's primary mirror. (Photograph courtesy of J. Hutchings, Dominion Astrophysical Observatory, and NASA)

a principal motivation for building and deploying this major space-based facility. Finally, many years later, Hubble did provide the clues astronomers needed in order to understand where and why these powerful beacons formed. The most widely accepted view now is that quasars are found in galaxies with active, supermassive black holes at their centers. Because of their enormous distance from Earth, the "host" galaxies appear very small and faint, and are very hard to see against the much brighter quasar light at their center.

1.2   THE HOST GALAXIES OF QUASARS

Observations with the Hubble Space Telescope, and more recently with newer ground-based telescopes that have the light-collecting ability to see to the edge of the universe, have established the fact that quasars reside in the nuclei of many different types of galaxy, from the normal to those highly disturbed by collisions or mergers. In all cases, however, the sites must provide the fuel to power these uniquely bright beacons. Later in this book, we will see that a quasar turns on when the supermassive black hole at the nucleus of one of these distant galaxies begins to accrete stars and gas from its nearby environment; the rate at which matter is converted into energy can be as high as ten Suns each and every year. The intense radiation field is emitted by the plasma on its final journey toward the event horizon – the threshold below which nothing can escape back into our universe. So the character and power of a quasar must depend on how much matter is available for consumption.

Most astronomers now believe that disturbances induced by gravitational interactions with neighboring galaxies trigger the infall of material toward the center of the quasar host galaxy. Large instruments, such as the Very Large Telescope on Paranal, Chile, can probe the environments of quasars as far away as 10 billion light-years or so, an example of which is shown in Fig. 1.4. In this case, the quasar (the bright point-like object at the center of the image) appears to be embedded within a complex structure consisting of several knots and arcs. In particular, the object just below the quasar image lies at a projected distance from the quasar of only 20 000 light-years, about two-thirds of the distance between the Earth and the center of the Milky Way galaxy. This object is most likely a companion that is interacting with the quasar host. Clearly, though, the most intriguing aspects of this image are the tails extending away from the quasar itself. Such tails are well known from nearby galaxy collisions, in which the gravitational fields of the participating structures tear at each other's fringes and produce the arc-like trailing debris. A spectacular example of this type of encounter is shown in Fig. 1.5, captured brilliantly in the southern constellation Corvus by the Hubble Space Telescope.

Many quasars, however, reside in apparently undisturbed galaxies, and this may be an indication that mechanisms other than a disruptive collision between such aggregates of matter may also be able to effectively fuel the supermassive black hole residing at the core. Prior to seeing the probing images produced by the Hubble Space Telescope, astronomers had reached the consensus that unless a collision is delivering large quantities of gas to the black hole, galaxies harboring these cosmic carnivores would supply fuel too slowly to sustain a full-blown quasar. In other words, whereas all quasars are supermassive black holes, not all supermassive black holes are visible as quasars. Instead, the astronomers thought, these gentler giants might sputter just enough to become the fainter galactic nuclei we see predominantly in our own galactic neighborhood. (A galaxy with a less active supermassive black hole than a quasar is called an active galaxy and its central massive core is known as an active galactic nucleus. Our Milky Way galaxy and our neighbor, the Andromeda galaxy, are examples of normal galaxies, where the supermassive black hole has very little nearby plasma to absorb. And, as we shall see in Chapter 4, some galaxies apparently do not have a supermassive black hole.) The question concerning how undisturbed galaxies spawn a quasar is still not fully answered. Perhaps the Next Generation Space Telescope, now under development and expected to fly in 2010, will be able to probe even deeper than the Hubble Space Telescope has done, and expose the additional clues we need to resolve this puzzle.

Over 15 000 distant quasars have now been found, following their early misidentification as unusual, nearby stars in the 1960s. We might wonder then why it is that quasars tend to shine from the edge of the visible universe, but seem to be completely absent in our vicinity where we tend to see predominantly the weaker (though sometimes active) galactic nuclei.

Because of their intrinsic brightness, the most distant quasars are seen at a time when the universe was but a fraction of its present age, roughly one billion years after the Big Bang. The current distance record is held by an object with the designation SDSS 1044-0125, which was discovered from data taken with the Sloan Digital Sky Survey, coordinated by the University of Chicago and the US Department of Energy's Fermi National Accelerator Laboratory. Conducting follow-up observations of this object with the unparalleled light-collecting power of the Keck Telescope in Hawaii to see its rainbow of colors and determine its redshift, a team led by Marc Davis at the University of California at Berkeley confirmed that it is indeed a quasar. They were surprised to learn that it is now the most distant object ever found in the universe, a beacon that must have been among the first objects ever to form. The light that we sense from it was emitted when the universe was almost ten times smaller than it is today, very close to the limit we should be able to see anywhere. This quasar is so distant that the expansion of the universe has shifted its light, originally emitted as ultraviolet photons, through the visible portion of the spectrum and into the infrared.

Astronomers sometimes refer to a telescope as a "time machine" because light travels at a finite speed, so that distant objects are seen as they existed in the past when their light was emitted. When we look at the Sun, we see it as it was eight minutes ago – the time it takes its radiation to reach us. Looking farther afield, we see nearby stars as they were several years ago, and when we marvel at the splendor of the Milky Way's neighboring galaxies, such as Andromeda, we are looking back in time several million years. Peering to the edge of the universe, where the first quasars ignited, we are therefore seeing activity that occurred over 10 billion years ago.

This link between the distance to quasars and the implied look-back time provides a clue we cannot ignore when we try to understand why these powerful objects appear to be absent in our local environment. The Sloan Digital Sky Survey has shown that the number of quasars rose dramatically from a billion years after the Big Bang to a peak around 2.5 billion years later, falling off sharply at later times toward the present.[5]

Quasars turn on when it appears that fresh matter is brought into their vicinity (see Fig. 1.4), and then fade into a barely perceptible glimmer not long thereafter. They apparently feed voraciously until their fuel is gone, so quasars and other types of active galactic nuclei, not to mention the relatively gentle giant at the heart of the Milky Way and in the core of Andromeda, are likely manifestations of the same phenomenon: a supermassive black hole gulping down hot gas at nearly the speed of light. Whether we see it as a quasar or as an active galactic nucleus probably has more to do with how much gas is present in its vicinity than anything else.

However, not all the supermassive black holes in our midst have necessarily grown through the quasar phase. As we shall soon see, quasars are objects compressing as many as one billion Suns within their girth, yet the black hole at the center of our galaxy is a svelte 2.6 million solar masses. Its neighbor in Andromeda is somewhat heavier, but not much more than about 10 million Suns. In other words, it does not look as though all the supermassive black holes in our vicinity are necessarily dormant quasars in the twilight of their lives. Indeed, we now have some evidence that the Milky Way and Andromeda are conspiring to create a quasar of their own one day from the building blocks now residing in their cores (see Chapter 4).

A rather remarkable recent discovery adds some credence to this story. Back in the late 1700s, the M82 galaxy got its name when it became the 82nd entry in a systematic catalog of nebulae and star clusters compiled by the French astronomer Charles Messier (1730–1817). Now, 220 years later, NASA's Chandra X-ray Observatory is

---

[5] When its work is completed, the Sloan project will ultimately have surveyed one quarter of the sky and 200 million objects. About 1 million of these will be quasars, which should provide a wealth of information for statistical and evolutionary studies.

zeroing in on what appears to be a mid-sized black hole located about 600 light-years from its center.[6] This object packs a mass of 500 Suns into a region no bigger than the moon. It is conceivable that this object might eventually sink to the center of M82, where it could then grow and eventually become a supermassive black hole in its own right, without having passed through the voracious eating phase that accounts for the quasar phenomenon.

These mid-mass black holes, a thousand times more massive than star-sized ones like Cygnus X-1 (of Walt Disney's "The Black Hole" movie fame), are beginning to define a class of their own. Yet they are still a thousand to a million times smaller than the largest variety, like the powerhouse in the core of 3C 273. Nonetheless, they behave very much like scaled-down versions of supermassive objects found in the nuclei of the most luminous galaxies, and continue to grow as they consume matter in their vicinity. This new category of objects suggests to us that not all the supermassive black holes in our vicinity must necessarily have begun their lives in catastrophic fashion during the quasar epoch. Some of them may have grown as malignant tumors on the substrate of existing galaxies.

1.3 THE ACTIVE NUCLEI OF "NORMAL" GALAXIES

Dazzling everyone with their display of raw power from literally the edge of the visible universe, quasars rightfully command our attention at the hierarchical peak of all the objects known to us. At the other end of the distance scale – within hundreds of thousands of light-years, as opposed to the 28 billion light-year expanse separating one edge of the universe from the other – our neighborhood is replete with giant black holes that may be just as massive as their quasar brethren, though their frugal eating habits prevent them from displaying the full range of activity we now recognize in the more distant objects. It seems that nature may be playing a cruel trick on those who dare to probe the mystery of supermassive black holes, since quasars are powerful, but too far away to study with precision (see, for

---

[6] For a detailed account of this discovery, see Matsumoto *et al.* (2001).

example, Fig. 1.2), while nearby, in the nucleus of Andromeda and in our own galactic center, the ponderous giants are starved and barely visible.

We shall learn in Chapter 6, however, that according to the latest results from the Hubble Space Telescope and the Chandra X-ray Telescope, as many as 200 million supermassive black holes may be lurking in the relative solitude of space. Couched in the nuclei of active galaxies, these objects thus occupy a very useful niche between the two extremes in distance. One of the most famous examples, Centaurus A, graces the southern constellation of Centaurus as a colorfully dramatic archetype of this group, only 11 million light-years away (see Fig. 1.6). Peering through the dark bands of dust toward the middle of this galaxy, the Hubble Space Telescope recently uncovered a disk of glowing, high-speed gas, swirling about a concentration of matter with the mass of 200 million Suns. It took a combined international effort, first with the Very Large Telescope at Paranal Observatory,[7] and then with an infrared detector on the Hubble Space Telescope (see Fig. 1.7), to finally reveal the culprit in a spectacle reminiscent of our discovery of 3C 273 in Fig. 1.2. The scientists who conducted these studies quickly realized that this enormous mass within the central cavity cannot be due to normal stars, since these objects would shine brightly, producing an intense optical spike toward the middle, unlike the rather tempered look of the infrared image shown here.

Centaurus A is an archetypical active galactic nucleus for another very important reason: it is apparently funneling highly energetic particles into beams perpendicular to the dark strands of dust. It may therefore have much in common with the X-ray jet-producing black hole in 3C 273 (see Fig. 1.2) and another well-known active galactic nucleus, Gygnus A, shown in Fig. 1.8.

---

[7] This observation was carried out by a team of astronomers from Italy, the UK, and the USA, including E. Schreier (Space Telescope Science Institute), Alessandro Marconi (Arcetri Observatory), Alessandro Capetti (Turin Observatory), David Axon (University of Hertfordshire), A. Koekemoer (Space Telescope Science Institute), and Duccio Macchetto (Space Telescope Science Institute).

It is generally believed that the intense radiation field from quasars and active galactic nuclei is produced when the infalling gas is compressed and heated to temperatures exceeding billions of degrees (see Chapter 2). Quasars and active galactic nuclei gobble up matter with such a ferocity that some spillage is unavoidable; radio and X-ray observations show that the jets of plasma screeching away at nearly the speed of light from the nucleus in Centaurus A and Cygnus A are not rare. As we shall see in Chapter 5, they form tightly confined streams of particles that blast through the galaxy and travel hundreds of thousands of light-years into intergalactic space.

In the next chapter, we will begin our pursuit of the exquisite clues astronomers have gathered regarding the most powerful objects in the universe, and of the comprehensive story they tell. The discovery of supermassive black holes – be they dormant behemoths in our galactic neighborhood, dark entities lurking in the nuclei of other more exotic galaxies, or even the powerful beacons at the edge of the visible universe – has stoked our sense of wonder and perplexed us with fundamental questions concerning the nature of spacetime and the origin and evolution of galaxies. How do these objects form and how does one "weigh" them? Are they really island universes whose interior is forever shielded from our view? While conducting such a scientific interrogation, astrophysicists often uncover more questions than they can answer. So be it. We shall push forward and expose what these bottomless pits in the fabric of space and time have to say about our origin and our ultimate destiny.

2 **Weighing supermassive objects**

Supermassive black holes are certainly the most powerful objects in the universe, yet even this attribution may not adequately convey the severity with which they stress their surroundings. Yes, their force of attraction is inexorable, but more than this, it is – as far as we can tell – infinitely unassailable once matter approaches so close that even something moving at the speed of light cannot break free. The radius at which this happens is known as the black hole's event horizon, for nothing within it can communicate with the universe outside. Thus, we have no way of directly seeing such an object. Instead, its presence may be deduced on the basis of the shadow it casts before a bright screen, such as a dense cluster of stars. To have any hope of carrying out such an observation, however, we must be close enough to the highly concentrated mass to actually resolve the dark depression among the myriad other details likely to be present in its environment.

We become aware of a supermassive black hole primarily because of the incomparable cosmic power it exudes. For example, the image of 3C 273 in Fig. 1.2 attests to its nature as one of the brightest beacons in the visible universe. Yet it should be black, drawing everything into a catastrophic fall toward oblivion, releasing nothing – particles or light – to breach its cloak of secrecy.

It turns out that supermassive black holes are luminous precisely because the material falling into them is squeezed and cajoled into producing radiation before disappearing forever below the horizon. They shine by proxy, inducing the hapless matter, trapped and moribund, to illuminate the otherwise engulfing darkness in the nascent universe. Much of what astrophysicists do when they study black holes is therefore concerned with the issue of what happens to matter as it descends into the precipice.

Recently, some very clever detective work by a group of astronomers in Nottingham and Birmingham has produced intriguing evidence that supermassive black holes have been absorbing mass relentlessly since the dawn of their existence, some 12 billion years ago. The difficulty they had to overcome in order to demonstrate this effect is due to the fact that black holes (and their host galaxies) have been around for a period far longer than one can measure on a human scale. Thus, the rate at which they acquire mass is imperceptible to beings with a lifespan 100 million times smaller than the age of the universe.

Rather than tracking the changes seen over time in a single object, however, one may instead compare the characteristics of black holes known to have different ages. Presumably the life history of each member of a given class follows more or less the same path, so that the mass difference between two objects with different ages must correspond to the mass gained by the older member over the intervening period.

The investigators from Nottingham and Birmingham[1] adopted this approach and assumed, in addition, that the age of a black hole within its host galaxy scales with the age of the galaxy itself. Galaxies grow old because the stars within them consume nuclear fuel. As the ashes sift through their interior, stars change color, so astronomers can tell how old a galaxy is by looking at its starlight. Thus far, this technique has produced a catalog of 23 nearby galaxies and their supermassive black holes, including well-known objects, such as the galaxy in Andromeda.

Looking at the trends exhibited by the members of this list, whose measured lifespan ranges from 4 to 12 billion years, the astronomers carrying out this study have found that the masses of black holes in young galaxies tend to be quite modest, while their counterparts in older galaxies tend to get progressively heavier with age. It thus appears that a supermassive black hole builds up its mass over

---

[1] See Merrifield, Forbes, and Terlevich (2000).

the entire history of the host galaxy, with no sign that the growth ever comes to an end. Its relentless weight gain is a consequence of the one-way flow permitted across the black hole's event horizon: stars and gas from the surrounding medium can be drawn in, but nothing can get out.

This story is interesting, you may say, and it constitutes an important piece of the puzzle, but it is somewhat premature because we have not yet explained how the mass of a black hole is actually determined. After all, that is the central focus of this chapter! The point of this account is to establish at the outset that a supermassive black hole is always accreting matter from its environment. We must also be aware that its feeding habit is the principal reason we know of its existence. Distant objects, certainly those at the edge of the visible universe, are too remote for us to study via their influence on the motion of stars around them. In many cases, we cannot even see the host galaxy or, at best, we can just catch its faint glimmer. But we do see the prodigious outpouring of energy radiated by the black hole and, like a radio or television signal transmitted through the Earth's atmosphere, the light that reaches us from this entity carries with it clues we can decipher; this light even carries information regarding the black hole's size and mass.

In several instances, the Hubble Space Telescope has provided us with direct visual evidence supporting our suspicion that supermassive black holes absorb matter continuously. A spectacular example is the dusty disk of material swirling about the nucleus in the elliptical galaxy NGC 7052 (i.e., the 7052nd entry in the New General Catalog), located in the constellation of Vulpecula, 191 million light-years from Earth (see Fig. 2.1). Appearing like a giant hubcap in space, this structure is possibly a remnant of an ancient galaxy collision, and will gradually be swallowed up by the black hole over the course of several billion years. The disk is redder than the light from the rest of the galaxy because dust absorbs blue light more effectively than red light, the same phenomenon that causes the Sun to redden toward sunset when its light must traverse a longer path through the atmosphere.

At the center of the disk, the bright spot is the accumulated light from stars that crowd around the black hole. In fact, this frenzy of stars locked tightly within the relatively small volume enveloping the center is itself an indication that a strong source of gravity must be lurking nearby, for otherwise they would disperse unharnessed and not produce the concentrated cusp that is so evident in this image.

2.1   ACCRETION OF PLASMA

The dusty disk orbiting the nucleus of NGC 7052 in Fig. 2.1, not to mention the tight clustering of stars in the very middle, demonstrates effectively how supermassive objects can easily overwhelm any matter in their vicinity. It may therefore seem odd to hear that astrophysicists actually have difficulty trying to understand how a black hole accretes its food. After all, haven't we just convinced ourselves that the pull of gravity is inexorable and eventually unassailable?

Matter avoids falling straight into a black hole for the same reason that the Moon does not fall directly to Earth, that cyclones form, and that an ice skater spins faster when she draws her arms in toward her body. It is the same reason why our Sun, thankfully, will orbit the center of the Milky Way galaxy at a safe distance (of 28 000 light-years) for an eternity – or at least until the Andromeda galaxy collides with us several billion years hence (see Chapter 4).

The guiding principle behind all of these phenomena is that sideways motion simply cannot disappear even if the pull directed toward a point in the middle is very strong. Any change in motion requires a force in that same direction. As the ice skater draws her arms in, she exerts a force toward her body, but the sideways motion of her arms is unaffected, and to compensate for the fact that the extent of her limbs is decreasing, her body spins faster. The Sun experiences a constant force of gravity toward the middle of our galaxy, but any influence in the direction of its motion along the circumference of its orbit is negligible. In the seventeenth century, Sir Isaac Newton (1642–1727) explained the motion of the Moon in this way. Earth's biggest satellite, he argued, is indeed falling toward us, but because it has so much sideways motion, it actually moves forward just as much as it moves downward. The net result is that it never quite reaches the surface of our planet, and instead continues to orbit in a circle for an indefinite period of time.

Similarly, the dusty gas in the disk surrounding NGC 7052 has so much sideways motion that even the unimaginable power of the central supermassive black hole in this system cannot immediately draw the matter inward. Something other than the black hole itself must first remove the sideways action. When galaxies collide, as we are currently witnessing with the Antennae in Fig. 1.5, the agitation caused by the turbulence created in the middle is quite sufficient to remove the sideways motion of the orbiting gas and thereby causes it to plummet toward the eagerly awaiting behemoth in the middle. This is the reason why quasars are believed to be the products of such galactic encounters and, at the same time, why the fact that some quasars apparently are not is difficult to understand.

Another recent piece of clever detective work, this time by a pair of astronomers at Ohio State University,[2] seems to have provided one of the first pieces of direct evidence that a host galaxy does indeed begin the inward transfer of matter to the central black hole from as far away as 1000 light-years or more – essentially from the outer edge of the dusty hubcap in Fig. 2.1.

When we look at the glistening spiral arms of the Milky Way, or the hauntingly beautiful cartwheel of the nearby Andromeda galaxy (not to mention the antennas of the colliding galaxies in Fig. 1.5), we are primarily looking at the patterns associated with starlight. But stars are difficult to shake from their orbits, particularly when they are far from the nucleus where the matter feeding the black hole must begin its inward journey. So knowledge of what the stars are doing cannot help us understand where the supermassive black hole finds its fuel. Instead, the most likely candidate is gas or, more accurately, the admixture of dust and gas that permeates the void between the stars.

[2] See Martini and Pogge (1999).

Using the Hubble Space Telescope, the Ohio State astronomers created images of the host galaxies in both visible light and near-infrared wavelengths. What they were looking for was the telltale signature of the radiation produced by dust rather than stars. Patterns in the dust would then also reveal the underlying structure of the gas, and finally answer the question of how the host galaxy feeds its nuclear master.

Radiation from stars is very easily absorbed and scattered by dust, but far less at near-infrared wavelengths than in the visible portion of the spectrum. So by combining the two views, the astronomers at Ohio State University could separate out the effects in their image due to starlight as opposed to those produced by matter in the interstellar medium. What they found was quite unexpected; they saw swirling patterns in the majority of galaxies, an example of which is shown in Fig. 2.2.

To understand the extent of the features evident in this photograph, imagine taking the dusty disk shown in Fig. 2.1 and turning it over on its side so that we are now looking at it face on. (The dimensions are approximately the same in these two views.) Though photographs such as this are meant to freeze the subject in time, the content of this particular image jumps to life as we recognize the whirlpool of dust and gas fueling the black hole in the middle. Unlike the much larger stellar spirals sweeping around the galaxy, this mini-spiral – 100 times smaller – appears to be directly connected to the central source of gravity.

Astronomers are quite animated about this novel result because it demonstrates in a very visual way how the excitation of spiral waves can produce an avalanche of material into the center. Like sound disturbances propagating through the air, these waves can carry, or take away, energy. And because the dusty gas swirling about the nucleus must pass through them, the interaction between the waves and the orbiting material can effectively reduce the sideways motion that would otherwise prevent the latter from falling directly into the black hole. Galaxies that display this beautiful pattern of dusty spiral waves

Figure 1.2 Discovered in the early 1960s, the quasar 3C 273 (the bright object in the upper-left-hand corner) was one of the first objects to be recognized as a "quasi-stellar-radio-source" (quasar), due to its incredible optical and radio brightness, though with very perplexing properties. Insightful analysis led astronomers to the profound conclusion that 3C 273, and the other members of its class, are not stars at all, but rather incredibly powerful objects billions of light-years away. This image was taken with NASA's Chandra X-ray Telescope, using its unrivaled high resolution. Its size is about 22 x 22 arcseconds, which at the distance to 3C 273, corresponds to roughly 2000 x 2000 light-years. Quasars often produce high-powered jets from their core at velocities very close to the speed of light. Here, Chandra is revealing for the first time the steady glow of X-rays from 3C 273's jet, each of whose bright knots is brighter than the entire high-energy output of many other similar objects. (Photograph courtesy of H. L. Marshall *et al.*, NASA, and MIT)

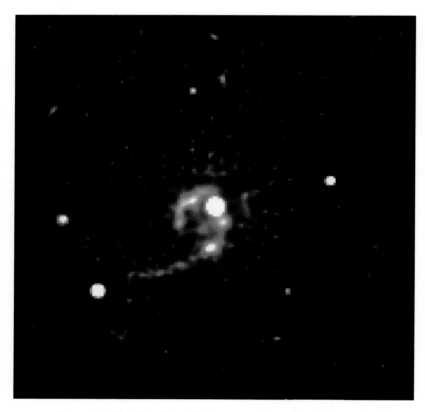

Figure 1.4 The Very Large Telescope on Paranal in Chile was used to obtain this spectacular image of the close and complex environment of the distant quasar HE 1013-2136. This object is seen in the southern constellation Hydra (the water snake) and is located some 10 billion light-years from Earth. The quasar is the point-like object at the center of this image. It is embedded within a complex structure consisting of two arched, knotty tails extending over more than 150,000 light-years, one-and-a-half times the diameter of the Milky Way. It is believed that the two tails result from a dramatic collision between the quasar host galaxy and one or more of the close companion galaxies. An example of such an encounter, occurring a mere 63 million light-years away from us, is shown in Fig. 1.5. (Photograph courtesy of Klaus Jäger *et al.* and the European Southern Observatory)

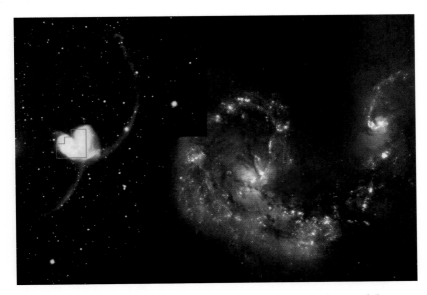

Figure 1.5 The left-hand image shows a ground-based view of the Antennae galaxies (known formally as NGC 4038/4039), located some 63 million light-years from Earth in the southern constellation Corvus. The pair of long, luminous tails were formed by the gravitational tidal forces between the two galaxies. The Hubble Space Telescope was able to zoom into the inner core region of this spectacular encounter, revealing a brilliant ``fireworks display'' with over 1000 bright, young clusters bursting to life in a sweeping spiral-like pattern traced by bright blue stars as a result of the head-on collision. The respective cores of the twin galaxies are the orange blobs, left and right of image center, crisscrossed by filaments of dark dust. A wide swath of chaotic dust stretches across the overlap region between the cores of the two galaxies. A collision such as this is thought to have formed the tails evident in Fig. 1.4. (Photograph courtesy of Brad Whitmore, the Space Telescope Science Institute, and NASA)

Figure 1.6 The constellation of Centaurus in the southern sky contains the nearest active galaxy to Earth. A mere 11 million light-years away, it is the brightest radio source in this region. This image is a composite of three photographs taken by the European Southern Observatory at Kueyen, Chile, using blue, green, and red filters. The dramatic dark band is thought to be the remnant of a smaller spiral galaxy that collided, and ultimately merged, with a large elliptical galaxy, not unlike the interaction currently deforming the stellar cartwheels in Fig. 1.5. Were we to image this hauntingly beautiful galaxy with a radio telescope, we would see two of the most spectacular jets of plasma spewing forth from the central region in a direction perpendicular to the dark dust lanes (see Figs. 5.7 and 5.8). These relativistic expulsions of plasma share much in common with the X-ray glowing stream we saw in Fig. 1.2, and with the radio-emitting beams of particles we will view in Fig. 1.8. (Photograph courtesy of Richard M. West and the European Southern Observatory)

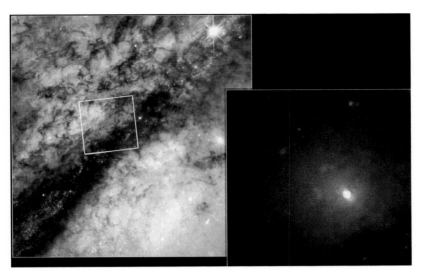

Figure 1.7 Peering through the center of the thick dust band in Centaurus A, Hubble's Near Infrared Camera Multi-Object Spectrometer (NICMOS) easily discovered a tilted disk of hot gas seen here as a (faint) white bar running diagonally over 130 light-years across the image center. The mass concentration required to corral the high-speed gas in the middle of this structure is inferred to be about 200 million Suns. The red blobs near the disk are glowing gas clouds that are being irradiated by the intense light from the active nucleus. The larger image to the left, whose central box is magnified at lower right, shows an enlarged view of the central region in Fig. 1.6, captured here with the Wide Field and Planetary Camera-2 (WFPC2) on the Hubble Space Telescope. (Photographs courtesy of E. J. Schreier at the Space Telescope Science Institute, and NASA)

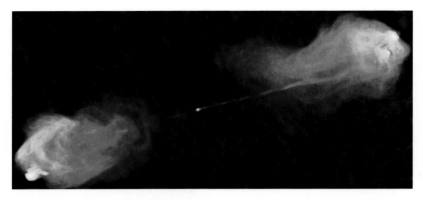

Figure 1.8 The Very Large Array (VLA) in Socorro, New Mexico, has produced many spectacular images of peculiar objects in the cosmos, but none more captivating than this glimpse of the powerful central engine and its relativistic ejection of plasma in the nucleus of the galaxy known as Cygnus A, in the constellation Cygnus (the swan). Taken at a radio wavelength of six centimeters, this glorious panorama reveals the highly ordered structure spanning over 500 000 light-years (more than three times the size of the entire Milky Way), fed by ultra-thin jets of energetic particles beamed from the compact radio core between them. The giant lobes are themselves formed when these jets plough into the tenuous gas that exists between galaxies. Despite its great distance from us (over 600 million light-years), it is still by far the closest powerful radio galaxy and one of the brightest radio sources in the sky. The fact that the jets must have been sustained in their tight configuration for over half a million (possibly as long as ten million) years means that a highly stable central object — probably a rapidly spinning supermassive black hole acting like an immovable gyroscope — must be the cause of all this activity. (Photograph courtesy of Chris Carilli, Rick Perley, NRAO, and AUI)

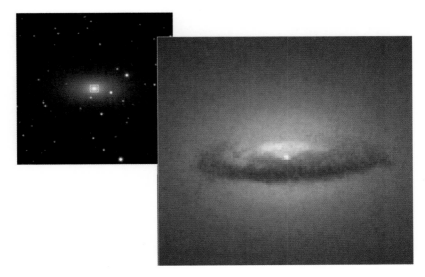

Figure 2.1 A giant (3700 light-year-diameter) disk of cold gas and dust encircles a 300 million solar-mass black hole in the core of the elliptical galaxy NGC 7052 (lower right). We see it in projection against the bright background screen of starlight; the front appears darker than the rear becasue the light from the other side of the nucleus must traverse across a great depth of material to reach us through the the edge of the disk nearest us. The dark, dusty disk represents a cold outer zone that extends inwards to an ultra-hot region. Gravity compresses and heats the gas, which then radiates at several wavelengths across the spectrum. The upper-left image shows the galaxy NGC 7052 as seen using a telescope from the ground, with the boxed region magnified at lower right. (Photograph courtesy of Roeland P. van der Marel at the Space Telescope Science Institute, Frank C. van den Bosch at the Universoty of Washington, and NASA)

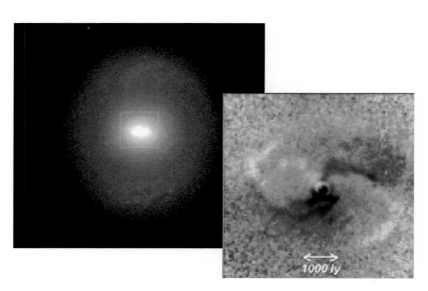

Figure 2.2 Using the Hubble Space Telescope to create an image of the galaxy known as Markarian 573 in both visible light and near-infrared wavelengths, astronomers have been able to merge the various features showing up in these views to produce this beautiful apparition of a dusty cyclone spiraling in toward the black hole in the nucleus of this host galaxy. The image at upper left shows the whole galaxy, with a central box to indicate the relative size of the magnified view to the lower right. About 100 times smaller than the cartwheeling spirals of stars we associate with galaxies such as the Milky Way and Andromeda, this miniature "mix-master" provides the necessary removal of sideways motion to permit the dust and gas to fall toward the center of the galaxy, igniting the supermassive black hole. (Image courtesy of Richard Pogge and Paul Martini of Ohio State University, and NASA)

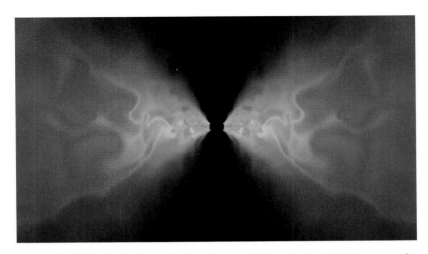

Figure 2.3 This is a simulated image of how the hot, infalling gas and dust would appear to us, were we to take a ride through the plane of the dusty disk in Fig. 2.1. However, before the compression and heating (which is revealed by the sequence of color changes toward the center of this view) would become apparent, we would need to approach the central object (realized here as a circular, dark depression) to within a distance roughly ten times the size of our solar system; by comparison, the extent of the dusty disk in Fig. 2.1 is about 3 million times the diameter of the solar system. The billowing clouds seen here in red suggest that much of the heat generated during the infall does not escape directly, but rather induces retrograde gas motions that sweep some of the plasma back out along the plane of the disk. (Image courtesy of Philip Armitage at the University of St Andrews and Kees Dullemond at the Max-Planck-Institut für Astrophysik)

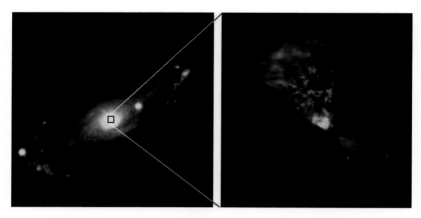

Figure 2.4 The left-hand panel shows a ground-based image of the barred spiral galaxy known as NGC 5728, located 125 million light-years away from Earth in the constellation Libra. Focusing on the nucleus of this active galaxy, the Hubble Space Telescope discovered a spectacular bi-conical beam of light emanating from the central supermassive black hole in this system (right-hand panel). It is believed that an orbiting ring of matter, like that simulated in Fig. 2.3, shapes the escaping ultraviolet radiation into two lighthouse beacons that ionize and excite the gas above and below the plane of an accretion disk feeding the central object. (Image courtesy of Andrew Wilson and collaborators at the Space Telescope Science Institute, the University of Maryland, Johns Hopkins University, Leiden Observatory, and NASA)

Figure 2.5 Peering into the nucleus of the peculiar galaxy NGC 4438, situated in the Virgo cluster some 50 million light-years from Earth, the Hubble Space Telescope caught the supermassive black hole in the act of blowing out a giant bubble away from the dark band of dust and gas spiraling toward the center. A second bubble can just barely be seen billowing downward below the dark swath of material. These hot bubbles are caused by the black hole's over-indulgent eating of matter trapped within the swirling accretion disk (seen as a white region below the bright bubble). The rate at which mass is spiraling inward is apparently so high that some of it is spewing out in opposite directions like high-powered garden hoses. (Image courtesy of Jeffrey Kenney and Elizabeth Yale at Yale University, and NASA)

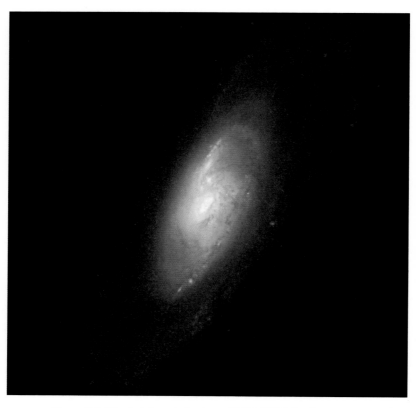

Figure 2.6 The bright spiral galaxy NGC 4258 (also known as M106) is receding from us with a speed of 537 kilometers per second, at a distance of about 23 million light-years from Earth. With an equatorial plane similarly inclined to that of the galaxy in Andromeda, many of its features resemble what we know about the Milky Way's sister galaxy. The dust lanes, for example, form a prominent spiral pattern that may be traced well into its bright central core region. Recently, the high angular resolution and sensitivity of the Very Long Baseline Array of the National Radio Astronomy Observatory allowed astronomers to carry out precise measurements of the motion of water masers within a light-year of the nucleus. The central mass implied by the structure in the disk containing these maser-emitting cloudlets is one of the most accurately measured to date, and has a value close to 40 million Suns. (Photograph courtesy of Zsolt Frei at Eötvös University and James Gunn at Princeton University)

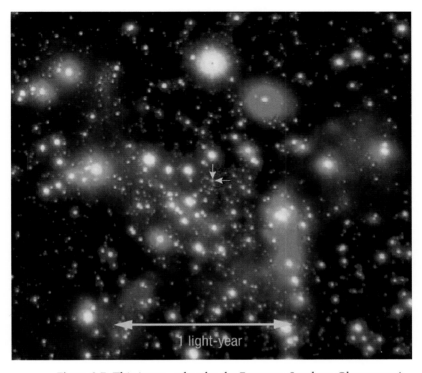

1 light-year

Figure 2.7 This image, taken by the European Southern Observatory's 8.2-meter telescope in Paranal, Chile, provides one of the sharpest views of the stars surrounding the supermassive black hole at the heart of the Milky Way. The colorization was produced by blending three photographs between 1.6 and 3.5 microns. The compact objects are stars and their colors indicate their temperature — blue is hot and red is cool — whereas the diffuse emission is produced by interstellar dust. The location of the black hole itself, which coincides with the center of the galaxy, is indicated by the two yellow arrows in the middle of the image. This view represents a scale of approximately 2 x 2 light-years, so the shadow cast by the black hole would be too small to see here. In addition, the black hole does not radiate perceptibly at infrared wavelengths, and therefore only the stars orbiting about the center are visible in this image. (Photograph courtesy of R. Genzel *et al.* at the Max-Planck-Institut für Extraterrestrische Physik, and the European Southern Observatory)

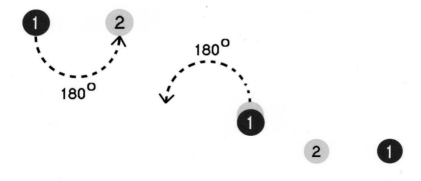

Figure 3.1 If an electron is occupying position 1, while a second electron occupies position 2, two individual rotations by 180 degrees are required to reverse their locations. The overall rotation is therefore 360 degrees, so that the orientation effect in three dimensional space becomes a factor in deciding whether or not the system is in the same state as the original.

Figure 3.2 The supermassive black hole at the nucleus of MCG-6—30—15, 130 million light-years from Earth, has the mass of about 100 million Suns. Recent observations with the XMM-Newton X-ray satellite indicate that some of the detected light was produced in a region very close to the black hole's event horizon, where the spacetime swirls in concert with the rotation of the central object. This forced motion entangles the magnetic field (shown in this artist's view as a wispy, bluish gossamer) anchored to the black hole and its surrounding accretion disk. (Image courtesy of the artist, Dana Berry, and SkyWorks Digital)

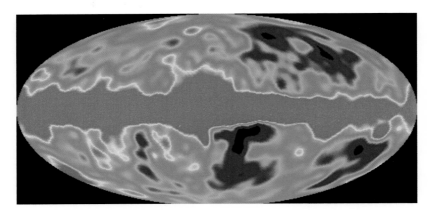

Figure 4.1 This map of the tiny variations in the temperature of the cosmic microwave background radiation shows where the fluctuations are located at the level of one part per million. The galactic center is located in the middle of this view, which spans the entire sky, from 90° to -90° in latitude (up and down) and from -180° to 180° in longitude (right to left). The earliest detection of this radiation, over 30 years ago, suggested that it was nearly uniform. But as instruments improved, features with finer resolution began to stand out until, in 1992, the Cosmic Background Explorer (COBE) satellite pierced the sky with sufficient resolution to grant astronomers the opportunity of producing this exquisite view. On this map, the hot regions, shown in red, are 0.0002 Kelvin hotter than the cold regions, shown in blue. Emission from the Milky Way galaxy dominates the equator of the map, but is virtually zero elsewhere. Away from the equator, fluctuations from the edge of the visible universe dominate the pattern of ``hot'' and ``cold.'' These cosmic microwave temperature fluctuations are believed to trace fluctuations in the density of matter in the early universe. Imprinted shortly after the Big Bang, they reveal the origin of galaxies and large scale structure in the cosmos. (Image courtesy of George F. Smoot *et al.* and NASA)

are therefore the likely hosts of very active supermassive objects in their core.

Imagine then that we could take a ride through this cyclone of dust and gas, approaching close enough to the supermassive black hole that we could begin to see the effects of strong gravity. Unfortunately, we cannot yet do this in reality, even by looking through a telescope. This is because the scale of the image in Fig. 2.2 is thousands of light-years, whereas we already know from the variability of quasars that the concentrated mass must be enclosed within a region no bigger than our solar system, which is over 3 million times smaller than the central dark spot in this figure! Astrophysicists can simulate such an adventure computationally, however, and with an appropriate choice of colors to represent the gas and its temperature, they can thereby create an image of how such an environment would appear were we to take this ride.

It turns out that we would need to get to within a distance of 10 to 100 solar-system diameters from the black hole before we could start to make out the significant compression and heating experienced by the infalling gas and dust. As we shall see in Chapter 3, on this scale the black hole itself would appear as a dark circular disk, roughly 100 times smaller. Thus, in the photograph shown in Fig. 2.2, the region of severe compression is well contained within the central black spot.

A beautiful example of the simulated image that can be created in this fashion is shown in Fig. 2.3, which illustrates how the infalling gas and dust would appear were we to approach the black hole – seen as a dark central orb – along the plane of the disk in Fig. 2.1. In other words, this image shows the density profile in a vertical cut through the "hubcap" in that figure. The color coding in this simulated view, which spans a region about ten times the size of our solar system, also demonstrates how quickly the gas is heated as it approaches the event horizon in the middle. The temperature can reach values as high as a million – sometimes a billion – degrees, so no solid structure would be able to survive this trip, turning instead into a hot tenuous plasma.

Astronomers know that the temperature can get this high because they sometimes see the effects of the radiation produced by the glowing matter. Looking at the simulated image in Fig. 2.3, imagine yourself gazing down at the black hole from above (or for that matter, gazing up at it from below). You would see most of the gas swirling about the middle, forming a conical opening in your direction. In contrast, the view we have in Fig. 2.3 is that of an observer in the orbital plane itself, and the conical openings point upward and downward in this image. The light produced by the hot compressed gas as it swirls about the black hole has an unobstructed escape route in your direction, but not so through the billowing matter. This effect reminds us of how a flashlight produces a directed beam of light, concentrating its emission in the forward direction, while preventing any leakage sideways through the opaque enclosure.

A quite remarkable *cosmic* version of this flashlight was discovered by the Hubble Space Telescope while focusing on the nucleus of the active galaxy NGC 5728 (see Fig. 2.4). The spectacular bi-conical beam of light emanating from the core in this object is ionizing the gas surrounding the center, and the red hues we see within the conical openings are produced by the irradiated plasma. To astronomers eager to learn about the nature of gas falling into a supermassive black hole, this discovery provides all the elements needed to confirm the anticipated existence of a funnel aperture in the central disk. The hot compressed plasma in this system glows with ultraviolet light, but a dense ring of gas and dust blocks Hubble's direct view of the black hole. Instead, some of this ionizing radiation is beamed into the open cones, aligned perpendicularly to us, out to several thousand light-years from the middle. The ring of matter effectively shapes the escaping ultraviolet light into two lighthouse beacons shining above and below the disk.

Something even more dramatic happens when the rate at which gas is spiraling into the black hole is too high to sustain a steady diet. The supermassive object's tendency is then to blow the excess matter and energy outward in the form of billowing bubbles of hot plasma (see Fig. 2.5). Acting like high-powered garden hoses, these twin ejections of material crash into a wall of dense gas and set it aglow. In this particular instance, the bubbles themselves are about 800 light-years long and 800 light-years wide, fitting easily within the central region of the dusty disk seen in Figs. 2.1 and 2.2. The bubble on the upper side of the dark bands in this image appears to be brighter because the expanding gas on this side is smashing into a denser clump of ambient matter. These bubbles are expected to expand until they eventually lose their energy, which should happen on a timescale much shorter than the age of the galaxy.

The theory of how supermassive black holes gain their mass is a fascinating subject in itself, and what we have seen here barely functions as an initial foray into this highly explored landscape. But let us now divert our attention to the question of how the radiation beamed in our direction by these imposing objects may be interrogated for clues that will ultimately reveal their proportion and mass. Viewed in retrospect, this achievement will seem quite remarkable, considering the fact that we are sensing the presence of these powerful entities from the other end of the universe.

2.2 DECIPHERING THE SIGNAL FROM THE INFALLING GAS

The attempt by Nicolaus Copernicus (1473–1543) to move the center of the universe from the Earth to the Sun, thereby creating the heliostatic cosmology, met with significant early resistance from the vast majority of his contemporaries because no one believed the Earth should be subjected to the implied distinct motions it must undertake in concert with the other planets. To a modern day reader, however, it may seem even more difficult to understand how Johannes Kepler's (1571–1630) explanation in support of this system, using the geometry of polyhedra, could have helped. In accounting for the fact that there were precisely six planets (the others had not yet been discovered, and the Moon had by then already been relegated to "attendant" status, which literally translated means "satellite"), Kepler suggested that:

> If a sphere were drawn to touch the inside of the path of Saturn, and a cube were inscribed in the sphere, then the sphere inscribed in that cube would be the sphere circumscribing the path of Jupiter. If a regular tetrahedron were then to be drawn in the sphere inscribing the path of Jupiter, the insphere of the tetrahedron would be the sphere circumscribing the path of Mars, and so inwards, putting the regular dodecahedron between Mars and Earth, the regular icosahedron between Earth and Venus, and the regular octahedron between Venus and Mercury.[3]

Miraculously, this description could not only account for the number of planets, but it also gave a convincing fit – within an error of 10 percent or so – to the sizes of the planetary paths derived by Copernicus. It all made much better sense when several years later Newton applied his revolutionary theory of gravitation to the motion of comets and planets in the solar system, to the Moon's orbit about the Earth, and to apples falling in Lincolnshire. Kepler's enduring legacy is the meticulous care he exercised while establishing the harmony of the heavens by charting the motion of our planetary neighbors as they wandered around the Sun. But it was Newton who proved once and for all that the planets must be moving under the influence of an inverse-square law force in order to obey all the empirical laws deduced earlier by Kepler.

Newton's law of gravitation actually has two principal components. In the *Principia*, he encapsulated these ideas as follows: "... all matter attracts all other matter with a force proportional to the product of their masses and inversely proportional to the square of the distance between them." Thus, knowledge of the planet's distance from the Sun, and the force that must be applied to it in order to keep its motion harnessed in perpetuity, is sufficient for physicists to

---

[3] For an extensive treatise on Kepler's contributions to cosmology, see Field (1987).

extract from Newton's law of gravitation the mass required to exert this influence.

Let us now take a significant leap forward, both in space across the vast intergalactic expanse to the far reaches of the universe, and in faith. If we assume that the laws of physics are the same "out there" as they are down here in our solar system – this is where the part about faith comes in – then Newton would argue that knowledge of the orbiting gas's velocity and its distance from the central black hole ought to be sufficient for us to infer the latter's mass. Thus, as long as we can discern these pertinent clues among the many details borne by the radiation beamed across the universe, we have at our disposal a cosmic balance with which to "weigh" a supermassive black hole.

Astronomers generally acknowledge that the most spectacular and compelling application of this technique has been made to the spiral galaxy known as NGC 4258, in the constellation Canes Venatici, not too far from the Big Dipper (also known as the Plough in some countries; see Fig. 2.6). An international team of Japanese and American astronomers used a continent-wide radio telescope, funded by the National Science Foundation, to observe a disk of dense molecular material orbiting within the galaxy's nucleus at velocities of up to 1050 kilometers per second.[4]

Drifting majestically some 23 million light-years from Earth, this spiral galaxy extends 90 000 light-years across, and is easily distinguished from the others via the microwave (maser) emission produced by water vapor in its nucleus. The acronym "maser" (microwave amplification by the stimulated emission of radiation) was coined by the Nobel Laureate Charles Townes and his collaborators in the mid twentieth century, after they successfully demonstrated the principle of microwave amplification in their laboratory. In this process, light of a certain frequency is amplified when it passes through a gas whose molecules are made to vibrate with a higher energy than normal. In NGC 4258, sufficient radiation is produced in the nucleus to excite

---

[4] See Miyoshi et al. (1995).

condensations of water molecules orbiting about the center, and this leads to strong stimulated emission at radio wavelengths. The disk within which these water molecules are trapped is tiny compared to the galaxy itself, but it is oriented fortuitously so that pencil-like beams of microwaves are directed toward us.

The swarm of photons produced in this fashion filter through the dust and gas enshrouding the central region without any attenuation, allowing us to probe the motion of matter deep within the core. Even so, these observations are not easily made. Radio astronomers must use a technique known as Very Long Baseline Interferometry (VLBI), which blends together the radio signals gathered simultaneously by many telescopes, often separated by many miles, or even thousands of miles. A major component in the arsenal of VLBI work is known as the Very Large Baseline Array (VLBA), which was commissioned in 1993. It constitutes a system of ten 82-foot-diameter dish antennas located across the USA from Hawaii to the Virgin Islands. All ten antennas work as a single instrument, controlled from an operations center in Socorro, New Mexico. For the observations of NGC 4258, the VLBA was joined by the Very Large Array (also used to generate the beautiful image in Fig. 1.8), a 27-antenna radio telescope in New Mexico.

These measurements of the maser emission from the nucleus of NGC 4258 tell us that water vapor is swept up by a giant whirlpool of material orbiting a strong source of gravity. In addition, the maser clouds appear to trace a very thin disk, with a motion that follows Johannes Kepler's orbital laws to within one part in 100, reaching a velocity of about 1050 kilometers per second at a distance of 0.5 light-years from the center. (The velocity itself is determined from the apparent wavelength shift of the radiation.) This is all we need to calculate from Newton's universal law of gravitation that approximately 35 to 40 million Suns must be concentrated within 0.5 light years of the center in NGC 4258.

The astronomers who made this groundbreaking discovery argue that the implied density of matter in this region must therefore be at least 100 million Suns per cubic light-year. If this mass were simply a highly concentrated star cluster, the stars would be separated by an average distance only somewhat greater than the diameter of the solar system and, with such proximity, they would not be able to survive the inevitable catastrophic collisions with each other. Because of the precision with which we can measure this concentration of dark mass, we regard the object housed in the nucleus of NGC 4258 as one of the two most compelling supermassive black holes now known, the other being the object prowling at the center of our own galaxy, about which we will have more to say in the following section.

However, the vast majority of quasars and active galaxies hosting supermassive black holes in their cores are not quite as compliant as NGC 4258. Taking into account the various factors that must all work in our favor, it is easy to understand why galaxies like NGC 4258 are indeed special and, therefore, rare. First, the conditions near the central object must conspire to maintain an excited population of water molecules, which facilitates the amplification of microwaves as they filter through the dusty gas. Second, the disk of infalling material must be relatively thin and oriented just right for us to see the thin beams of maser light skimming along the plane of this disk. And finally, the host galaxy must be sufficiently close to us that we can resolve several maser-emitting cloudlets for a proper determination of the distances involved, as well as the velocities. The number of host galaxies to which this application may be made is therefore understandably small.

Well, astronomers are a clever lot and, though not as accurate as the maser method, several alternative techniques have been developed to provide information on the speed of matter and its distance from the central black hole, whose mass may therefore be extracted with comparable validity using the Newtonian "balance." Let us take another look at Fig. 2.4, and imagine that the cosmic flashlight effect we are witnessing here is functioning with equal or better efficiency even closer to the supermassive object in the center of its host galaxy. That is, we should envisage a situation in which clouds of gas orbiting the nucleus are being irradiated by the central beacon and that, in turn, they glow with hues we can recognize and identify in terms of specific atomic wavelengths.

Then, just as Maarten Schmidt could confidently argue in favor of 3C 273's fantastic speed of recession, we would recognize the shifts in wavelengths produced by these clouds as being indicative of their velocity. By now, this is no longer viewed as an unusually clever measurement, but the method used to determine the distance from the black hole to the glowing gas, known as reverberation, is rather inventive.

By monitoring the light emitted by the supermassive black hole and, independently, the glow produced by its halo of irradiated clouds of gas, astronomers can sense when a variation in the radiative output has occurred. It so happens that when the quasar varies its brightness, so does the surrounding matter – but only after a certain time delay. The lag is clearly due to the time it took the irradiating light from the center to reach the clouds and, since we know how fast photons can travel, this delay provides a measure of the distance between the nucleus and the orbiting plasma. And so, with the speed and distance known, we again have recourse to Sir Isaac Newton and his law of universal gravitation to extract the quasar's mass.

2.3   THE CENTER OF OUR GALAXY

The supermassive objects we have discussed thus far exhibit black-hole activity in spectacular ways. Echoing their presence with unmitigated power from early in the universe's expansion, for example, quasars are difficult to supplant as the most unusual entities in existence. But size is not everything. Known as Sagittarius A*, our very own black hole at the center of the Milky Way may not be the most massive, nor the most energetic, but it is by the far the closest. And what a difference a few million light-years can make!

Looking at a photograph of a galaxy such as NGC 4258 (Fig. 2.6), one could understandably be duped into thinking that this aggregate of stars is packed together rather tightly. Yet over most of a galaxy's extent, stars account for an infinitesimal fraction of its volume. For example, if we were to think of a star as a cherry, we would need to commute between the major cities in Europe to simulate a typical distance (of several light-years) between stellar neighbors in space. At the center of our galaxy, however, some 10 million stars swarm within a mere light-year of the nucleus.

The brightest members of this crowded field are captured in Fig. 2.7, an infrared photograph of unprecedented clarity produced recently with the 8.2-meter VLT YEPUN telescope at the European Southern Observatory in Paranal, Chile.[5] The image we see here is sharp because of a technique known as adaptive optics, in which a mirror in the telescope moves constantly to correct for the effects of turbulence in the Earth's atmosphere. This motion of the air produces a twinkle in far-away objects and distorts and blurs their appearance on photographs such as this. Adaptive optics can in principle create images with a clarity that is even greater than that of the Hubble Space Telescope, whose primary mirror has an aperture three times smaller than that of the VLT YEPUN.

The laws of planetary motion deduced by Kepler and Newton dictate that objects move faster the closer they orbit about the central source of gravity. Mercury, for example, the planet closest to the Sun, completes one orbit every 88 Earth-days, whereas Pluto, meandering about the farthest reaches of the solar system, takes a full 90 465 Earth-days to accomplish the same feat. This is understandable, of course, in terms of how quickly the Sun's gravitational pull diminishes with distance – this is the whole point of the Newtonian "balance" we have been using to weigh supermassive black holes in the nuclei of their host galaxies.

Sagittarius A* is so close to us compared to its brethren else-where in the universe, that on an image like that in Fig. 2.7 we can identify individual stars orbiting a mere seven to ten light-days from

[5] Each of the four telescopes in the Very Large Telescope array has been assigned a name based on objects known to the Mapuche people, who live in the area south of the Bio-Bio river, some 500 kilometers from Santiago de Chile. YEPUN, the fourth telescope in this set, means Venus, or evening star.

the source of gravity.[6] In the nucleus of Andromeda, the nearest major galaxy to the Milky Way, the best we could currently manage is about two light-years.

With this proximity to the supermassive black hole, stars orbit at blistering speeds of up to 5 million kilometers per hour, allowing us to see their motion in real time – even at the 28 000 light-year distance to the galactic center. They are zipping along so fast, in fact, that astronomers can easily detect a shift in their position on photographic plates taken only several years apart. Their incredible rate of advance makes it possible now to unambiguously trace their orbits with startling precision, revealing periods as short as 15 years! Compare this with the 220 million years it takes the Sun to encircle the galactic center just once.

The most spectacular identification[7] to date of a star orbiting about the black hole was announced in October 2002 by an international team of astronomers using the unparalleled light-gathering capability of the VLT YEPUN telescope that produced the stunning image shown in Fig. 2.7. If one looks closely at the middle of this photograph, it appears that one of the fainter stars – designated as S2 – lies right on top of the position where the black hole is inferred to be. S2 is an otherwise "normal" star, though some 15 times more massive and seven times larger than the Sun.

That in itself is not very surprising, since a chance coincidence in the projected position of two objects along the line of sight is not unusual in a crowded field such as this. What is amazing, however, is that the star S2 has been tracked now for over ten years and the loci defining its path trace a perfect ellipse with one focus at the very

---

[6] This effort has benefited from the contributions made by several observatories around the world. The principal investigators leading the effort to use these techniques for imaging the galactic center have been Andrea Ghez, Mark Morris, Eric Becklin, and their collaborators at UCLA, and Reinhard Genzel, Andreas Eckart (now at the University of Köln), and their collaborators at the Max Planck Institut in Garching, Germany.

[7] This discovery was reported in *Nature* (2002) by a large team of astronomers led by Schödel, at the Max-Planck-Institut für Extraterrestrische Physik.

position of the supermassive black hole. This photograph, taken near the middle of 2002, just happens to have caught S2 at the point of closest approach (known as the perenigricon, as opposed to the point farthest away, known as the aponigricon), making it look as though it was sitting right on top of the nucleus.

At this position, the star S2 was a mere 17 light-hours away from the black hole – roughly three times the distance between the Sun and Pluto, while traveling with a speed in excess of 5000 kilometers per second, by far the most extreme measurements ever made for such an orbit and velocity. Indeed, when this photograph was taken, the astronomers realized that the star S2 had just performed a rapid swing-by near the center, creating an unprecedented opportunity of determining not only the precise position of the source of gravity, but also its strength, and thereby its mass.

Using Newton's universal law of gravitation, we infer that the mass required to harness the motion of stars such as S2 seen closest to Sagittarius A* at the galactic center is *2.6 million Suns*, compressed into a region no bigger than about seven light-days, and possibly just 17 light-hours given this latest discovery. For this reason, Sagittarius A*, and its cousin in the nucleus of NGC 4258, whose maser-emitting disk betrays its 40 million solar-mass heft, are considered by astronomers to be the most precisely "weighed" supermassive black holes thus far discovered.

In the coming chapters, we will probe more deeply into the nature of these objects and why they are increasingly being viewed as fundamental building blocks of structure in the universe. It appears that as much as half of all the radiation pervading the void of space may have been produced by these megalithic entities. Some were here near the dawn of time, perhaps even before galaxies as we know them formed, and all will be here toward the end.

3 The black hole spacetime

Settling on the banks of the Tiber river, the Latini would establish a city in the seventh century BC that later came to dominate much of the civilized world. They used the word *gravis* to denote heavy or serious, and the corresponding noun *gravitas* for heaviness and weight. Our modern word *gravity*, and its more precise derivative *gravitation*, trace their roots to this early usage, which itself is linked to a yet older root that includes the Sanskrit *guru* (for weighty or venerable), among others. The ancients were evidently quite aware of this ever-present property of matter – that it should have an unwavering attraction toward the Earth – though up to the time of Galileo and Newton, gravity simply remained a name for the phenomenon, without any explanation or even an adequate description.

3.1 THE INEXORABLE FORCE OF GRAVITY

Toward the end of the seventeenth century, attempts to account for the behavior of objects changing their motion in response to external influences were primarily concerned with the nature of forces that one could easily identify. In the story of Goliath's slaying, for example, the stone was dispatched toward his forehead after David released the sling. Prior to that moment, the diminutive combatant was able to restrain the motion of the stone with a force applied by his hand mediated through the string. Newton argued that the Earth must itself be exerting an attractive force on matter since everything falls down in the same direction. This action seemed to be quite unfathomable, however, since there was no obvious way in which the Earth could reach out to pull objects toward it. The influence evidently had to be produced by some "action at a distance," in which one object feels the presence of another just by virtue of its existence at a different location. Newton understandably viewed gravity as being special, since it appeared to be distinct from other forces, like the hands that pushed, or the chains that pulled, and he found this to be very troublesome.

We now know, of course, that this distinction is only an illusion created by the different scales over which the forces act. The forces exerted by strings and hands are macroscopic manifestations of a microscopic equivalent of gravity's "action at a distance," except that in this case the action is due to electromagnetism. The reason this book does not fall through your hands is that there are repulsive forces between the charged particles in each of them that prevent the former's downward motion. But this electrostatic recoil is mediated across the space between the constituent particles in much the same way that gravity acts between the Earth and the Moon – the electrons and other atomic constituents never really "touch" at all.

Newton's anxiety about the status of gravity as an action-at-a-distance influence was certainly well placed, though not for the reasons he envisaged, because modern physics has adopted a considerably different view of how forces are exerted on one particle by another. The illusory distinction between gravitational and electromagnetic influences is no longer relevant and, later in this chapter, we will theorize that gravity and all other forces are in fact merely different manifestations of a single, unified action.

The classical theory of gravitation, in which one object pulls on another with a force proportional to their masses and inversely proportional to the square of their separation, represents one of the most significant triumphs in science. Its predictions have been confirmed by observations with astounding accuracy. John Couch Adams (1819–92) in England and Jean Joseph Le Verrier (1811–77) in France, for example, used slight peculiarities in the motion of Uranus to predict in 1846 the existence of Neptune, one of the first major successes of this theory.[1]

---

[1] The story of Neptune's prediction and eventual discovery highlights some of the difficulties faced by early researchers dealing with inefficient means of communication, poor coordination, and modest facilities. This tale is told by Moore (1996), among others.

As far as we can tell, Newton's law of gravitation applies to objects without any limit on their separation, though physicists do not know quite yet what to make of new evidence that the universe's expansion has been speeding up rather than slowing down. A straightforward consideration of this physical law would indicate that since all matter attracts all other matter, the ensuing "inwardly" directed gravitational force should slow the expansion down. Yet looking farther and farther into the past, by examining the motion of the most distant objects visible with the largest telescopes, astronomers are finding that the young universe was expanding at a speed lower than expected.

It is still too early to tell what the eventual outcome of this discovery will be. Some theorists are proposing that space contains a hitherto unseen energy field (a "dark" energy) whose pressure counteracts the effects of gravity, much as the air pumped into a balloon can make it expand even if other forces (e.g., the elasticity of its rubber) conspire to keep it deflated. Others believe this accelerated expansion is evidence for an odd form of matter that exhibits gravitational repulsion, rather than attraction, which would then clearly require a modification to Newton's law of gravitation.[2] But even such a change, if needed, would have an inconsequential impact on the description of gravity on a galactic scale or smaller. So Newton's simple inverse-square law appears to be a good starting point for any discussion concerning the nature of supermassive black holes.

Having said this, it is nonetheless true that certain aspects of gravity are to this day profoundly mysterious on all scales. Wherever we look in the heavens, its influence is overpowering and interminable. Yet accustomed to the quick fix, fast food, and life in the express lane, we cannot help but notice the languorous pace with which the universe succumbs to change. The cosmos projects a majesty derived in part from its lethargic vastness, and it is specifically gravity's duplicitous character – always the dominant force, yet weak – that troubles physicists deeply.

---

[2] Some recent ideas on this topic may be found in Ostriker and Steinhardt (2001).

On the surface of the Earth, for example, something may break free and fall to the ground with quite a thud. But if you were to put a nail on a tabletop, the introduction of even just a toy magnet into the system would be sufficient to "rip" it away from Earth's clutches. Whereas it takes our whole planet to hold the nail in place, the electromagnetic force exerted by the comparatively tiny magnet is much, much stronger.

When scientists shift their attention away from the cosmologically large to the microscopically small, probing deeply into the structure of an atom, or its nucleus, the effects of gravity are completely insignificant compared to the forces that bind the individual particles together. On this scale, the structure of matter is entirely beholden to the influence of non-gravitational forces, whose incredible strength produces a compression of the constituents into relatively small volumes. Thus, it happens that an atom is much smaller than a galaxy because the electromagnetic force that couples the orbiting electrons to the central positively charged nucleus is many factors of ten stronger than the mutual gravitational attraction between stars swirling within the galactic aggregate.

On large scales, this astounding disparity in strength produces electrically neutral clumps of matter. Charged plasma simply cannot remain isolated due to the overwhelming electric forces it experiences from its surroundings. Thus, gravity is dominant, at least on a planetary or galactic scale, because it happens to be the only remaining influence at such distances. This part, at least, is easily comprehensible, but what puzzles physicists is something more subtle than the variation in size – it has to do with the fact that when size is prescribed by gravity, it is way off-scale compared to that associated with any of the other forces. In other words, even though the other forces may differ in strength relative to each other, gravity is uncommonly weak, a situation known as the "Hierarchy Problem."

To put this in more concrete terms, the gravitational attraction between two electrons is weaker by a factor of 10 followed by 42 zeros when compared with the force of repulsion they experience due to their mutual electric force. A pair of electrons would each have to be ten thousand billion billion times more massive in order for the electric and gravitational forces between them to be equal. (Incidentally, the resource required to produce such a heavy particle is known as the Planck energy.)

So why do we need to worry about all this? Well, for one thing, the behavior of matter probably changes its character considerably when it executes a transition from a relatively diffuse structure the size of a galaxy or a star, down toward the apparent singularity of a black hole. Eventually, astrophysicists will need to understand better than they do now what in fact happens when matter collapses to such a degree that the fundamental forces readjust their relative strengths. For example, suppose that on very small spatial scales some hitherto unforeseen factor causes gravity to bridge the "hierarchical gap" and merge with the other forces – perhaps even become unified with them. Will this then preclude the existence of singularities?

Not surprisingly, some of the most sparkling work in theoretical physics now is directed toward the problem of how we should include gravity in a complete description of all the fundamental forces of nature. And one of the most appealing threads in this discourse hinges on the possible existence of extra spatial dimensions, beyond the three we encounter in our everyday experience.

3.2 UNSEEN DIMENSIONS

In a nutshell, this hypothesis holds that gravity only appears to be weak because electromagnetism and the other forces are constrained to function solely within our familiar three-dimensional space (plus time), whereas gravity is free to influence activity in all dimensions. Thus, what we experience in our three-dimensional world is but a fraction of gravity's overall strength.

The very mention of extra dimensions is bizarre to physicists and non-physicists alike, though the historical pedigree for such a concept continues to mature. The word "dimension" tends to be used in many contexts, often lacking consistency with its true meaning. A dimension grants us the access to move in a certain direction without changing our distance from another given straight line drawn in this space. This may sound very abstract, but in actuality humans construct habitable environments that adhere to this rule – and bring out the three-dimensional structure of our world – quite explicitly. Let us imagine entering a room whose shape is typically that of a rectangular box, often simply a cube. Now begin at one corner of that room (say with your right shoulder brushing up against the wall for definiteness), and slide along that wall toward the neighboring corner. What you should notice is that as you do this, your distance from the wall on your left side remains fixed. The distance from your feet to the ceiling also does not change.

Starting at one corner of the room, you can choose precisely three different and independent directions in which your motion will produce this same result. But proceed in any other direction, say the diagonal, and your distance from the walls will change with every step. This cubical room therefore encloses a well-defined, three-dimensional space.

Not all spaces need to have exactly three dimensions, however. Ants crawling along the surface of a large sphere inhabit a two-dimensional world, because they can move in precisely two independent directions while preserving a constant distance from given axes, these being embedded in the surface itself. The ants trapped in this environment do not have the luxury of moving along a radius anchored to the center of their spherical world, which would have constituted the third independent direction – essentially the sense of up and down in our three-dimensional room.

This particular type of space is interesting for several other important reasons, including the fact that it is infinite, yet bounded, and multiply connected. An ant crawling along this surface can continue on its journey forever without reaching an end; it may retrace its path an infinite number of times by going around the entire circumference of the sphere, but there is never an end to its path. Nonetheless, mathematicians say it is bounded because the actual amount of area covered by the surface is a finite number. This space is also multiply connected because any given point on the spherical surface may be reached by taking many – actually an infinite number of – paths.

Taking this reduction in the number of dimensions one step further, it is evident that a space comprised solely of a circle represents a one-dimensional world, since an ant crawling in this environment has precisely one direction in which it can move. Again this space is infinite, though bounded (because of the same property we ascribed to the surface of a sphere), and it is multiply-connected in the sense that the ant may reach any given point on the circle by moving either clockwise or counterclockwise in its highly restricted world.

With this background in mind, the question that filters through the imagination of modern-day physicists and mathematicians is whether nature "chose" to have us function in a three-dimensional space because that's all there is, or whether we might be the three-dimensional analog of the ants crawling on their two-dimensional surface. Since we can easily envisage spaces with one and two dimensions, there is no *a priori* reason to assume that three is the ultimate number.

The idea that there may exist a four-dimensional hyperspace (as it is called), and that our three-dimensional space is its surface – just as the ants' world is the two-dimensional surface of the three-dimensional volume enclosed within their sphere – was first described in Georg Riemann's (1826–66) doctoral thesis in 1851. But it appears that such a notion was taken seriously only after a then-unknown Prussian mathematician, Theodor Kaluza (1885–1954), submitted an intriguing proposal for a four-dimensional space to Albert Einstein. By introducing one additional dimension, he argued in a letter written in 1919, the gravitational and electromagnetic fields could be unified into a single entity, which only manifests itself as different components when we project it back into our space. Struck by its significant originality, Einstein sponsored Kaluza's article for publication in 1921.

Shortly thereafter, one of Kaluza's former students, Oskar Klein (1894–1977), postulated in 1926 that the fourth spatial dimension might be curled up so tightly that it would not be detectable in subatomic packets of space. This development has historical significance because Kaluza-Klein bottles – as this compact extra dimension was later known – became the foundation for modern-day string theory.[3]

String theory has now become one of the most active branches of theoretical physics, claiming perhaps as many as half of all active researchers whose primary interest lies in the nature of particles and their interactions. It is a description of reality in which the tiniest entities are very short strings floating through spacetime, rather than the point-like fabrications of the more established (though perhaps now outdated) scientific disciplines. In this theory, the most likely number of spatial dimensions is 10, yielding a total of 11 when time is included. As it turns out, though, an attempt to find an explanation for gravity's extreme weakness relative to the other forces does not necessarily have to be wedded to the notion of strings.

The reason for this is that the existence of extra dimensions, regardless of whether particles are points or strings, will result in the rapid dilution of the gravitational force away from the source. A significant surge in interest for these ideas followed the publication of a provocative paper in 1998, in which particle theorists Nima Arkani-Hamed, Savas Dimopoulos (both then at Stanford university), and Gia Dvali (then at Trieste), proposed the existence of two or more spatial dimensions curled up not on a subatomic scale, as mainstream string theory presumes, but rather up to a scale as large as a few millimeters.[4] Such enormous hidden dimensions could in principle lead to detectable departures from Newtonian gravity with a new generation of more sensitive tabletop experiments.

To understand how the strength of gravity depends on the number of dimensions, let us consider the following simple analogy.

---

[3] A relatively non-technical description of string theory may be found in Brian Greene (2000). For an introduction to Kaluza-Klein theories, see also Lee (1984).
[4] See Arkani-Hamed, Dimopoulos, and Dvali (1998).

Imagine a dozen people huddling together, all facing away from the group. If the huddle forms a tight circle, each must be gazing in an outwardly radial direction. They begin to walk, maintaining a circular shape as they go. Not surprisingly, the separation between each pair of individuals increases as the circumference of the circle grows. The point is that whereas the number of people never changes, the circle's perimeter gets bigger as they recede from its center.

Not wanting to waste this effort, we ask our experimenters to sweep the floor as they go. But we notice that as the circle grows, they find it more and more difficult to keep up with their chore because each individual's region grows in proportion to the distance he covers.

If we could somehow position our sweepers on the surface of a sphere, and carry out a similar thought experiment for this type of space, we would now find that the region associated with each of them grows even faster than it did before, because the area increases as the *square* of the distance they cover. By adding one dimension, we have caused the density of people to dilute even faster as they recede from the center.

On a three-dimensional surface within a four-dimensional hyperspace, each sweeper would need to keep up with a region that grows in proportion to the *cube* of the distance he covers. Obviously, this trend continues indefinitely as we add more and more spatial dimensions. Gravity's influence, represented here metaphorically by the sweeping potency of our workforce, also ebbs with increasing distance away from the source. But the critical point is that the surface "area" of our space scales with a progressively higher power of the distance as the number of dimensions grows. So the rate of diminishing influence is greater, the larger the number of dimensions.

The payoff from this somewhat abstract reasoning comes next when we try to couch these various ideas into the language of real numbers and consequences. We are struck first by the fact that the number of dimensions needed to solve the Hierarchy Problem depends on how large each of them is. This we know: the Newtonian prescription for gravity – that it scales inversely as the square of the distance between two objects – holds true for planets in the solar system, for stars orbiting about the center of the galaxy, and apparently also for the expansion of the universe. Thus, on these scales at least, there must be precisely three spatial dimensions since the power with which the strength of gravity dilutes – the square of the distance in this case – is one smaller than this number.

For such huge distances, we also know that the influence of gravity is disquietingly small compared to that of the other forces in nature. Perhaps the reason gravity is so weak on large scales, argue the proponents of extra dimensions, is that the dilution takes place much closer to the source. In order for this to work, however, the extra dimensions must be curled back on themselves with a size no bigger than current measurements allow. Taking the simplest approach – adding a single new spatial dimension – is therefore already ruled out, because its radius would have to be roughly the distance between the Earth and the Sun in order for gravity's strength to dilute sufficiently.

Two extra dimensions can solve the Hierarchy Problem if they are about one millimeter in extent – precisely where our current direct knowledge of gravity ends. At the time of writing, a dozen or so groups are eagerly looking for departures from Newtonian gravity (which scales as the inverse square of the distance) at these small, yet macroscopic, scales. A provisional result was reported recently by one of these outfits at the Marcel Grossmann Meeting on General Relativity in Rome. Representing the University of Washington's Eot-Wash laboratory, Jens Gundlach revealed that the group's specially designed torsion balance has yet to uncover any evidence for extra dimensions down to a distance as small as 0.2 millimeters.[5]

The extra dimensions are even smaller if there are more than two, and if string theory turns out to be the correct description of nature, its seven additional spatial dimensions (beyond the three we inhabit) would be no bigger than the size of a uranium nucleus. That's why astronomers are not concerned about these fundamental changes

[5] See Gundlach and Merkowitz (2000).

to gravity, since a dramatic increase in its strength at these short distances would not affect objects held together on planetary, stellar or galactic scales.

Interestingly, though, even a size as small as a uranium nucleus *will* become accessible to physicists with the next generation of particle-collider experiments. By smashing particles together with a thousand times more energy than is contained within such a nucleus, the Large Hadron Collider at the CERN laboratory in Geneva, Switzerland, may, by 2010, begin to see evidence for a significant leakage of gravitational energy out of our three-dimensional space during such an encounter. Physicists anticipate this loss if the proponents of extra spatial dimensions are correct, and gravity's strength ramps up dramatically at this predicted distance. A more unsettling hypothesis holds that gravity's added influence on these small scales may in fact lead to the proliferation of micro black holes which, however, would be too small to cause us any problems, since they would radiate all of their energy and evaporate in a time too small to even measure with current instruments.

How fortunate our descendants will be to know the dimensionality of our existence and, by extension, perhaps even to unveil other universes that may exist in parallel to our own a mere millimeter or less away! Astrophysicists generally agree that a key ingredient in the development of this understanding will be our resolution of the supermassive black-hole mystery. Given sufficient time, even a weak gravity will eventually cause matter to collapse to densities at which the effects described in this section will manifest themselves. Let us therefore pursue this thought and take guidance from what current physical laws have to tell us about the ultimate fate of matter tumbling into the oblivion of its own event horizon.

3.3 MATTER'S FUTILE RESISTANCE TO TOTAL COLLAPSE
Since gravity is the dominant force acting over large distances, its inexorable pull should evidently always lead to strong condensations of matter. Can anything ever stop it?

The answer, of course, is yes – albeit temporarily. The Sun, for example, much heavier than our planet, has managed to sustain itself against a terminal collapse for over 4 billion years. Aside from the support provided by sideways motion that we discussed in Chapter 2, particles cannot slide effortlessly through or past each other, so when collisions between them become very frequent, any further compression is halted, or at least postponed.

Think of it this way. People filtering into a large ballroom experience very little resistance initially because of the large space available to them. As the density in the crowd increases, however, collisions between individuals eventually prevents any tighter compression. Once all the bumping and squeezing forces those present to fill the entire room, the influx stops, and the final number of guests fitting into this space depends on how lively they are. A dormant, placid crowd will be very dense because their motion is suppressed, whereas an enthusiastic, energetic group of individuals will enliven the space with activity that spreads the individuals apart. As the evening progresses, however, the guests start to tire and as the ballroom's collective ebullience diminishes, so too does the outward push from all their collisions and motion. Those waiting outside the doors will eventually be able to filter in as the crowd condenses farther and farther toward the middle of the floor.

Within a clump of matter, the outward support against catastrophic gravitational collapse is first provided by the heat generated by compression (in the same way that a bicycle pump heats up when we squeeze the air inside it), and then by the burning of nuclear fuel – a process that converts hydrogen into helium, and then carbon, nitrogen, oxygen, and so forth down the chain toward heavier elements. Even in a giant ball of gas, however, there's a limit to how much heat can thus be generated, since eventually all the fuel will have turned into iron, which will simply not burn in a thermonuclear reaction. The core accumulates ashes, growing larger and larger with the passage of time. Matter loses its first major battle with gravity when its heat is gone, and it collapses to even higher densities. Yet rather than ease its hold, gravity continues to push the growing behemoth toward the precipice of total collapse.

But something quite astonishing happens when the density of matter becomes so large that the electrons flooding the medium start to overlap with each other's positions. The effect has to do with the marriage of a truly remarkable property of our three-dimensional space with the tenets of quantum mechanics. The basis for this unexpected phenomenon – and the reason why we are even here to consider it – is the observation that when an object is rotated by 360 degrees, it returns to a state that *looks* the same as before geometrically, but is in fact quite distinct with respect to its surroundings. A second full rotation (for a total of 720 degrees) is needed to bring the object back to its original state.

An elegant way to prove this statement is the following. Take a small cube and paint each of its sides with a different color to distinguish it from the others. Then place the cube in the center of a larger box and use a string to connect each of its eight vertices to the corresponding internal corner of the enclosing box. Leave enough slack so that you can rotate the cube and maneuver it when the strings get twisted. Now rotate the cube through 360 degrees. The threads become tangled. Nothing you can do will unwind them and you will find it impossible to arrange every string in such a way that it extends from vertex to corner along a straight line.

Rotating the cube once more about the same axis by another 360 degrees, you will find that the threads appear to become even more twisted. But this is only an illusion, for the cube and its entanglement with the box are now actually back in a state identical to that at the beginning of the experiment. Amazingly, it only takes a little effort moving the strings around, while keeping the cube perfectly still, to completely straighten them out. After a rotation by 720 degrees, the threads run as they did in the beginning, forming straight lines between the vertices of the cube and the corners of the bounding box.

This result may be generalized to other rotations, but only in multiples of 720 degrees. For example, a total rotation by 1440 or 2160 degrees will also leave the cube unaffected in orientation and twisting relative to its surroundings. Evidently there is something about the geometry of orientation in three-dimensional space that is not fully taken into account when we consider just the cube's aspect, since without the strings, the cube appears to be in the same state as it was in the beginning even after a rotation of only 360 degrees.

Fortunately, there is an even simpler demonstration – known as the Philippine wine dance – that everyone can use to visualize this effect if the experiment with the cube is not doable.[6] Bend and orient your right arm so that your elbow points upwards, letting your forearm and hand dangle to the ground. Then bend your wrist so that the palm of your right hand is facing up. Place a glass – the dextrous among us will want to fill it with wine first – on your palm, and slowly rotate your hand 360 degrees in a plane parallel to the floor. Your right hand and the glass should now appear to have the same orientation they had originally, but your forearm is twisted and your elbow points to the ground. If you continue to rotate the glass in your hand by another 360 degrees (completing this Cleopatra-like movement), it will recover its original aspect and position. Remarkably, instead of incurring an additional twist, your forearm rediscovers its original shape and orientation!

This fundamental property of three-dimensional space has quite a profound influence on the behavior of matter and our existence in general. In nature, there are two principal categories of particles – those for which this effect is irrelevant and others for which the impact of this phenomenon manifests itself in very visible and important ways. And the cause of this is the fact that the sideways motion of matter we talked about in the previous chapter is quantized when

---

[6] For a technical description of this and other demonstrations of the twisting of space, see Bernstein and Phillips (1981).

circumstances force it to be confined within atomic-sized regions of space. For example, if we could take a spinning top and shrink it to arbitrarily small sizes, nature would only permit it to have certain discrete rates of spin and nothing in between.[7]

The factor that distinguishes one category of particles from the other is whether or not their spinning can be projected to zero when viewed relative to some pre-selected direction. Particles for which the answer to this question is yes are known as bosons; the rest are called fermions – in honor of the two legendary physicists, Bose and Fermi. A trivial example of the former is a particle that has no spin at all, for then regardless of our perspective, its projected spinning action is always zero.[8] Some particles, like electrons, protons, and neutrons – the principal components of an atom – always have spin, and possess it with certain discrete values that no matter what orientation we select, the projection of that spin is never zero.[9]

Thus emerges the nexus that weds the tenets of quantum mechanics to the remarkable property of orientation in three-dimensional space. The swirling motion of our diminutive spinning top and the spin of elementary particles are the "threads" that entangle these objects to their surroundings. Particles whose spin can be projected to zero are untethered, and a rotation by 360 degrees restores them to their exact original state. This is the situation we would have encountered with our multi-colored cube were we to cut all the strings connecting its vertices to the corresponding corners of the bounding box. Fermions, however, are always tethered since they have a spin that can be sensed from any orientation in the surrounding space. For these, therefore, a rotation by 360 degrees would leave

[7] It's somewhat dangerous to use this analogy since the spin of an elementary particle is evidently not associated with the swirling motion of something tangible and extended. Perhaps that motion is taking place in small, extra dimensions not directly viewable in our three-dimensional world.

[8] We will leave out of this discussion particles – such as photons – that have no mass and travel at the speed of light. For them, the spin behavior is different yet again.

[9] This too arises from the quantization of sideways motion imposed by quantum mechanics, since fractional values of the projected spin are not supportable by nature if they fall within the finite number of allowed discrete values.

them in a state that is different and easily identified as separate from the original!

The second tenet of quantum mechanics bearing on this story holds that two particles of the same type are completely indistinguishable if they have the same energy and other dynamical properties. If we were to place two electrons adjacent to each other, we could not under any circumstance label them in such a way that we could then keep track of which is where. Thus, to describe this situation, we must allow for all possibilities – that the first electron is in the left position and the second is on the right, *and* that the first electron is on the right, while the second is on the left. And this is precisely where we run head first into the peculiar problem of orientation in three-dimensional space.

The reason is the following. Once we know that there are two electrons occupying adjacent positions, then we also know that each of these particles may be manifested in either of the two locations because of their indistinguishability. But a geometry consisting of the first electron on the left with the second on the right is rotated by exactly 360 degrees relative to one in which the first electron is on the right and the second on the left. Since the two electrons are disconnected, it is not enough to simply rotate one of them by 180 degrees about the other; the second needs to be independently moved to the location where the first used to be. So together, these two individual rotations by 180 degrees produce an overall rotation of 360 degrees (see Fig. 3.1).

This transformed configuration, however, is *not* identical to the original one because of the orientation effect described above. In quantum mechanics, the fact that the same system of two particles must simultaneously occupy two distinct states therefore results in an abrogation of this configuration as one of the allowed equilibria for the system!

There's only one way to fix this – and that is to endow the electrons with properties that allow them to occupy distinguishable states. In other words, when particles such as this come sufficiently close together that they start overlapping each other's position, something about them must be different, otherwise they are simply not permitted to exist together. That something might be their energy, or it might be the degree of sideways motion that they have relative to the nucleus in the atom. For bosons, this is never an issue since a rotation of 360 degrees does not change their state. Thus, nature permits any number of them to co-exist, whether or not they are distinguishable. In recent years, for example, physicists have found ways to create something called a Bose-Einstein Condensate. In simple terms, this means that a configuration of particles is established in which a very large number of bosons all condense down to a very low-energy state – the *same* state.

But this is not possible for fermions – the principal constituents of atoms, molecules, cells, and life. Whenever these gather in numbers, they must be distinguishable, and that means they must have different energies and sideways motion relative to a force center. This is precisely the reason why the electrons in atoms must occupy stratified levels, and why the various elements therefore occupy different positions in the periodic table. Chemistry, biology – indeed, our very existence – would not otherwise have substance. And all of this is due to that peculiar property of orientation in three-dimensional space.

Matter collapsing under its own weight must eventually reach a density at which this restriction on the co-existence of identical fermions becomes an issue. This happens when the condensation of mass loses its heat and electrons begin to occupy neighboring states. But their density cannot be so high that nature runs out of low-energy *distinguishable* slots in which to place them. Any collapsing ball of gas has a finite amount of energy, and if the density increases indefinitely, some of the electrons must eventually fill slots that are no longer accessible. At that point, the implosion stops.

Millions of objects like this, known as white dwarfs, dot the galaxy. If the Sun were to collapse to such a configuration, it would be no bigger than the Earth. The Sun is actually close to being the biggest star that can survive catastrophic collapse because of the support provided by its electrons.

But what happens if there is so much matter around that it continues to pile up on top of this compact core? There is no evidence at all that quantum mechanics can break down. In fact, quantum mechanics is one of the most resilient theories in science, perhaps because in comparison with any of the other attempts at explaining nature, it is more of a description than a theory. Instead, when an additional half to two solar masses of material has fallen onto such an object, the electrons and protons are squeezed together and fuse to form a new class of fermions known as neutrons and, losing its support from free electrons, the star resumes its implosion. The ensuing collapse is halted once again – apparently the last – when the issue of distinguishability resurfaces, though this time it is the neutrons that cannot all occupy the same state. Astronomers call such an object a neutron star. It is so compact that a Sun's worth of material is squeezed into a sphere the size of Chicago, and its density is so off-scale that a single teaspoon full of neutron-star matter weighs as much as all of humanity combined.

As we shall see in the next section, neutron stars hover precariously above the size limit imposed by general relativity. The Sun, for example, would have to shrink to roughly 3 kilometers in radius in order to become a black hole. As a neutron star, it would have a radius no bigger than 10 to 20 kilometers. So what happens next as matter continues to pile on and gravity maintains its relentless pursuit toward total collapse? After all, these objects are comparable in mass to the Sun, but the supermassive objects in the nuclei of galaxies and quasars are millions to billions of times heavier.

The physics of matter under such extreme conditions starts to get somewhat murky at this point, given that astrophysicists have very little to guide them. It is known that as the mass of a neutron star increases, its size diminishes, but the exact rate at which this happens is still uncertain. By the time its mass has increased to about three or four Suns, its radius may be as small as 7 or 8 kilometers, and the gravitational pull on its surface is so strong that even light struggles titanically to escape. Indeed, for a three-solar-mass object, this radius is already below the event horizon, and we would call the "new" object a black hole.

Anything we would say about the behavior of matter beyond this state is pure speculation, because we have absolutely no information at all about what transpires across the "membrane of no return." If quantum mechanics survives intact, the neutrons will never slide together toward a singularity, though they probably would dissolve into a plasma comprised of their constituents – quarks. But these too are fermions, and would presumably therefore still be subjected to the exclusion principle arising from their indistinguishability at close range. It may happen that for some reason fermions transform into bosons under such circumstances. We simply don't know.[10] Quantum mechanics and the classical theory of general relativity diverge drastically here, because on the one hand, quantum mechanics does not permit particles to infiltrate into each other's space, whereas general relativity seems to suggest that progress toward a singularity is unavoidable. What we do know, however, is that in either case the compressed material is entombed below an event horizon, and we have no way of communicating with it farther, unless we too take the irreversible plunge.

3.4 THE BLACK HOLE SPACETIME

A significant fraction of this theorizing is based on a remarkable insight displayed almost a century ago by a follower of Einstein. In the December twilight of 1915, Karl Schwarzschild (1873–1916) was stationed on the Russian front when he received copies of Einstein's papers outlining the theory of general relativity. He unfortunately contracted an illness soon thereafter, and died upon his return to Potsdam, where he had been a professor before the outbreak of war in

---

[10] Some discussion on this topic, which is still valid today, may be found in the pioneering papers by Chandrasekhar (1931), Landau (1932), Eddington (1935), and Oppenheimer and Snyder (1939).

1914. Quite remarkably, he was able within that very short window of opportunity to derive a solution to the equations of general relativity, describing the behavior of space and time surrounding a spherical mass. Schwarzschild sent his paper to Einstein, who would transmit it to the Berlin Academy, and received the following response: "I had not expected that the exact solution to the problem could be formulated. Your analytical treatment of the problem appears to me splendid." Nonetheless, neither of the two men could yet appreciate how significant this contribution would be, for it contained a complete description of the external *metric* (i.e., a mathematical prescription for measuring distance and time) of a spherical, electrically neutral, nonrotating black hole, which today is often called a "Schwarzschild black hole," in honor of the volunteer soldier who first found a way to describe it while fighting the debilitating cold on Europe's eastern flank.

Understanding the structure of spacetime really amounts to knowing how distances and intervals of time are viewed by one individual relative to another. Everything else – be it the velocity of a particle falling toward a source of gravity or its acceleration in an elevator – is derived from these. Nature reveals to us that light possesses the highest attainable speed (even the effects of gravity cannot propagate faster) and that the distances and times we measure are altered between frames in such a way as to preserve this speed everywhere and always. We don't yet understand why this happens, but physicists do have the ability to produce a metric to interpret these measurements.

In special relativity – which does not include any acceleration and is therefore only an appropriate description of spacetime in the absence of gravity – the metric is very simple to write down and to understand. Since light advances at 300 000 kilometers per second, it will have traveled a distance of 300 000 kilometers times the number of seconds that have elapsed in a given interval. So regardless of which person is making the observation, the measured distance for a light pulse will be $d = ct$, in which c is the speed of light and t is the interval of time in seconds. Since a light pulse can either be moving toward us or away from us – two cases with opposite signs in d – it is often more convenient and conventional to write $d^2 = (ct)^2$, that is, the square of the distance traveled by light is the square of its speed times the square of the elapsed time. This relationship is the metric of special relativity in the sense that no matter who is making the measurement, the observer will conclude that their value of d^2 for light is always given by c^2 times their value of t^2.

In the presence of gravity, however, the very act of letting time advance produces a transformation into a frame moving even faster than the one before it. Schwarzschild theorized that we cannot simply have $d^2 = (ct)^2$ since the final measurement for the determination of d occurs in a different frame than that of the initial measurement. He recognized that a modification was needed in order to preserve the constancy of c, and concluded that this must be a factor that depends on the local value of the gravitational acceleration.

Schwarzschild's metric for the propagation of light in the case of a static, spherically symmetric source of gravity is $d^2/f = (ct)^2 f$, in which f is defined to be the factor $(1 - 2GM/c^2r)$, M is the mass of the central object, r is the radius from its center, and G is a constant characterizing the strength of gravity. The proof that this metric actually satisfies Einstein's equations requires some mathematical effort. Understanding its physical meaning, however, is not that difficult.[11] One can see right away that the effects due to the gravitational acceleration enter in a very simple way. Newton's monumental effort in formulating the law of gravity may be recognized (albeit in a modified form) within the factor f. In fact, the term $2GM/c^2r$ is the ratio of the Newtonian escape speed (from an object with mass M) to the speed of light, c, all squared.

---

[11] Note that this form of the metric is correct as long as light is moving along the radius, meaning that d here is to be viewed as a measure of the change in r only. If the ray of light is moving in any other direction, the correction due to the acceleration appearing on the left-hand side is then set equal to one. However, the correction associated with the passage of time (appearing on the right-hand side) is always present.

Looking at Schwarzschild's metric, we see that far away from the object, where the radius is large, the factor f approaches unity, which recovers the metric of special relativity, $d^2 = (ct)^2$. This limiting behavior is realized because the force of gravity decreases inversely as the square of the radius, so if the distance is large enough, the influence of *any* object becomes enfeebled. But when we view what happens to light going the other way, toward the source of gravity, something very unusual occurs as the magnitude of the escape velocity approaches the speed of light. The factor f goes to zero, and the radius, $r = 2GM/c^2$, at which this unfolds is known as the *Schwarzschild radius*.

Particles reaching this level or closer can never escape back into our universe, because they would have to move faster than light to extricate themselves from the clutches of the central object. The spherical surface at the Schwarzschild radius therefore divides the exterior universe from the inaccessible, interior region. Someone finding himself just outside of this surface can potentially move to larger radii and escape without having to attain light speed. But below this surface, even light cannot escape. The interior region is consequently black, and objects that produce a spacetime with a Schwarzschild radius reachable by infalling matter (or light) are called *black holes*, and their "surface" is known as the *event horizon*.

Thinking back to the previous section, we can now understand why neutron stars are such intriguing objects. Digesting a Sun's worth of mass within their 10-kilometer girth, these objects boast a Schwarzschild radius, $2GM/c^2$, of about 3 kilometers, and hovering tantalizingly close to the point of no return, they are primed for any upheaval associated with the additional collapse of fresh material that would drive the entire system into oblivion. Astrophysicists believe that this occurs when two more Sun's worth of matter have fallen in. The Schwarzschild radius will by then have grown to 9 kilometers, engulfing the anticipated highly compressed size of the new star. After an event horizon forms, it hardly matters what else accretes toward the dark object. Scaling as $2GM/c^2$, its Schwarzschild radius only gets bigger as more mass pours in, eventually reaching solar-system proportions when a billion Suns will have funneled into the pit.

None of the other issues we have explored previously – the existence of extra dimensions or the indistinguishability of like particles – can prevent this from happening, though it is an enduring mystery why all the "physics" associated with the collapse of matter and the eventual formation of its event horizon takes place before the object has grown to no more than a handful of solar masses. After all, the gap in mass between objects of this size, and those throbbing with the power of a billion Suns at the edge of the visible universe, is quite enormous. But at least we now recognize why this disparity exists – it is yet another manifestation of the Hierarchy Problem.

3.5   ROTATING BLACK HOLES

Schwarzschild's remarkable achievement would have retained its pivotal significance even to this day, had it not been for another breakthrough solution to the equations of general relativity that eventually eclipsed it. Fifty years after the first formulation of spacetime in a gravitational field, Roy Kerr's impassioned pursuit of a more appropriate spacetime surrounding a *rotating* black hole finally paid off.

Most objects in the universe spin at least a little because it is virtually impossible to assemble an aggregate of matter with components that move radially inward. Sideways motion is so prevalent in nature that it often functions as a powerful diagnostic of the forces that shape the various concentrations of mass (see Chapter 2). To form a nonrotating star, all of the sideways motion in the gaseous crucible where it condenses must somehow be removed. Otherwise, that sideways motion would imprint itself on the compressed object, just as a spinning figure skater speeds up when she pulls in her arms.

This phenomenon can produce furiously spinning entities when their progenitors collapse into ultracompact volumes. Should our Sun approach old age as a white dwarf, shrinking by a factor of 100 in size, its current rotational cycle of 26 days would shorten into a period of only 4 minutes. And if it were to condense a thousand times more – cascading into a neutron star – it would then rotate several hundred times per second.

Black holes are born spinning. They grow with the additional acquisition of matter throughout their existence, and absorb whatever sideways motion filters across their event horizon. Astronomers believe that eventually everything near this surface of no return may be moving sideways at nearly the speed of light. Even a black hole with the mass of 100 million Suns and a circumference stretching over 150 million kilometers, could be rotating with a period of only one hour and three-quarters!

Schwarzschild's metric cannot handle this because a rotating black hole impacts the spacetime around it in unexpected and challenging ways. It is no longer possible to say that $d^2/f = (ct)^2 f$, for the same reason that a mother watching her son jumping onto a merry-go-round cannot measure the total distance he covers by simply counting his steps. Though she may still be able to monitor the passage of time t by tracking the ticks on her watch, the actual distance d that her son travels is now augmented by the merry-go-round's sideways motion, which carries him along for the ride.

Roy Kerr's solution to the equations of general relativity shows that the spacetime itself, like water in a whirlpool, swirls around the black hole with a speed wedded to the latter's spin, though decreasing with distance from the center. Physicists call this effect "frame dragging," meaning that the spacetime itself and all its contents are forced into co-rotation with the source of gravity, even if objects in that frame are completely stationary relative to the space itself. This is quite a bizarre concept, and difficult to accept at face value, for it seems to imply that even if we could somehow place a particle with zero sideways motion in the vicinity of a spinning black hole, the fact that the underlying spacetime is rotating means that the particle would still appear to be moving sideways from the perspective of a distant observer.

The phenomenon of a rotating spacetime is very interesting indeed and may be the cause of several peculiar characteristics of supermassive black holes. Evidently, this spinning effect may actually make it easier for matter to resist the inward pull of gravity, since it has more sideways motion to lose before succumbing to its inevitable collapse. In addition, the rotation axis provides a virtually permanent anchor to which everything else in quasars and active galactic nuclei may be referenced. Another cursory look at Figs. 1.2 and 1.8 convinces us that the highly entrained stream of plasma ejected from the center of each object represents a very stable, long-lived activity. For a supermassive black hole, the most secure dynamic in the long run is its axis of rotation, and these magnificent ultra-thin plumes are nature's way of showing us the direction of that spin.

The recent launch of several powerful X-ray satellites is providing astronomers with additional compelling evidence for the existence of rotation in black holes. One of these missions, a spaced-based X-ray telescope with a multiple-mirror design – hence its early designation as the XMM facility – was launched from French Guiana by the European Space Agency (ESA) and NASA in December 1999. Carrying three advanced X-ray components with the light-collecting ability to detect millions of objects, XMM-Newton (as it is now called) is far more sensitive than any previous X-ray instrument. Already, it has richly rewarded investigators using it with the discovery of what appears to be a rapidly spinning, 100-million-solar-mass black hole in the heart of MCG-6–30–15, a galaxy 130 million light-years away.

From Earth, MCG-6–30–15 doesn't look particularly unusual – it hasn't even been endowed with a recognizable name, like the Milky Way, or Andromeda. It is a lenticular, lens-shaped aggregate of stars lacking the eye-pleasing glistening spiral arms gracing our own galaxy. But XMM-Newton sees something else; peering into the nucleus of this structure, it detects the X-rays produced by hot, glowing gas squeezing into the black hole.[12]

---

[12] This work was carried out by an international team of astronomers using the ESA/NASA satellite XMM-Newton (see Wilms *et al.*, 2001).

X-rays are much more energetic than visible light, and are often produced when particles collide or are otherwise heated to very high temperatures. For example, dentists can create an image of your teeth using X-rays emerging from a machine that sprays a metal target with a beam of accelerated particles. The ensuing melée produces a pinball cascade of electrons trapped inside their atoms as they jump from one level to the next, shedding radiation with every bounce. The excited – though trapped – electrons in an iron target sometimes lose a precise 6.4 kilo-electron volts of energy with their jump. Something very similar happens with the gas swirling about the black hole, except that the projectiles are now very energetic photons crashing into free-floating iron atoms. The result is the re-emission of X-ray light with the same precise energy of 6.4 kilo-electron volts, easily detectable with XMM-Newton.

But astronomers do not see these photons right at 6.4 kilo-electron volts because the iron atoms in MCG-6–30–15 are moving. That is actually even better, because the X-rays are Doppler-shifted, like a radar beam reflecting off a speeding car. The radar waves return to the monitoring device more frequently if the car is moving toward the policeman; less often if it is receding. The X-radiation arriving at Earth from MCG-6–30–15 is therefore shifted toward the blue end of the spectrum (i.e., to higher energies) and intensified on the side of the accretion disk (see Chapter 2) that is moving toward us; it is red-shifted toward lower energies on the side that is moving away.

The signal that is reaching us now from MCG-6–30–15 left its source during Earth's early Cretaceous Period, some 130 million years ago. In June 2000, some of it entered the open hatch at one end of XMM-Newton, glanced off highly polished gold mirrors, and came to a focus onto a silicon wafer at the other end of the spacecraft, 25 feet away. These X-ray photons carried with them some very startling information that was quickly deciphered by the investigators, though even they could not believe what their early analysis was telling them.

The X-ray glow of those iron atoms, they found, is so intense that compression by gravity alone could not possibly explain it. Moreover, the X-ray photons are Doppler-shifted to such a degree that they must have originated much closer to the center of the black hole than expected simply on the basis of the Schwarzschild metric. Only the additional support from the sideways motion induced by a rotating spacetime could allow the iron atoms to get so close to the event horizon without being catastrophically drawn in. The supermassive black hole in MCG-6-30-15 must evidently be rotating very rapidly. And if these atoms are radiating with uncommon brightness, something other than gravity must be heating them.

For several decades now, theorists have been wondering whether it may be possible to tap into the rotational energy of the swirling spacetime and convert it into some of the universe's most spectacular displays – the enormous outpouring of light seen from quasars (see Fig. 1.2) and the jets of radiant gas that shoot out of certain active galaxies at near light speed. The basic principle behind this was actually discovered in 1831 by the English chemist and physicist Michael Faraday (1791–1867). Faraday had a broad range of interests, including the condensation of gases, metallurgy, optical illusions, acoustics, and the conservation of energy. Many of his discoveries are considered to have been of groundbreaking importance: electrostatic induction (1838), the relationship between electricity and magnetism (1838) and between electricity and gravity (1851), hydroelectricity (1843), and atmospheric magnetism (1851).

In 1831, he discovered induced current in his best-known experiment, where a galvanometer showed the existence of current in a coil wrapped around a current-carrying metal ring. This is the principle underlying what was later to be known as the transformer. Later that year, he approached his electric motor from the other direction, hypothesizing that a moving magnet could produce an electric current, thereby creating the first dynamo (or generator).

The accretion disk surrounding the black hole in MCG-6-30-15 contains charged particles that generate a magnetic field when they move. In 1977, Roger Blandford and Roman Znajek of the University of Cambridge proposed that the infalling particles and their magnetic field lines plunge into the black hole together, so these lines protrude from the event horizon like quills on a porcupine. But this is just like the setup employed by Faraday in his experiment, albeit on a much grander scale, with a magnetic field moving through an electrical conductor. The spinning black hole induces a rotation in the surrounding spacetime and moves the magnetic field lines around it, creating voltages (see Fig. 3.2). Unlike Faraday's tabletop experiment, however, these voltages can be prodigiously large – the voltage difference between the poles of the black hole and its equator may be billions of trillions of volts.

This looks like a paradox at first because energy seems to be extracted from the black hole, but the resource tapped by this cosmic generator was never actually swallowed in the first place. Rather, it was stored in the spacetime whirlpool outside the event horizon. The magnetic field lines function as wires in an enormous electric circuit that thread this rotating region, rendering the black hole itself the generator. It may be thought of as an enormous flywheel, slowing down ever so slightly as the magnetic field lines fling electrically charged particles into distant space. Like rubber bands, they get twisted up and then snap back, repeating this cycle over and over again, producing a pulsating ejection of particles and energy.

Some of this pyrotechnic display is probably what ends up as narrow streams of energized gas that emerges from the cores of certain galaxies at more than 99 percent the speed of light, penetrating several million light-years into intergalactic space, and then splattering into giant luminous lobes (see Fig. 1.8). The rest of the liberated energy presumably heats up the gas surrounding the black hole, producing the uncommonly bright iron X-ray emission detected from MCG-6–30–15 by XMM-Newton.

The curious thing right now is why MCG-6–30–15 itself does *not* have any discernible jets. For an active galaxy it is relatively quiet. It does, however, have an unusually bright ring of X-ray emitting iron around its nuclear girth. As we have seen, the most plausible origin for this light is Faraday's electromagnetic generator powered by the spinning black hole. Perhaps there exists a "valve" not yet recognized that regulates how much of the tapped energy is partitioned into the outpouring of energized particles and the local heating of orbiting gas. We will no doubt hear more in the coming years from astrophysicists working on this fascinating problem.

4 Formation of supermassive black holes

Though some Hubble images of distant galaxies feature destructive collisions that could trigger quasar activity, others show that many normal, undisturbed aggregates of stars are oblivious to the cosmic thunder within their midst. This is an indication that a variety of mechanisms – some quite subtle – may be responsible for igniting a quasar. Whatever the formative process is, however, these supermassive objects seem to have spared their hosts from any obvious damage, so their prodigious outpouring of matter and radiation may be a short-lived phenomenon.[1] Still, this observation is not sufficient to guide astronomers toward the identification of a coherent, single pattern of quasar birth and growth.

For years, astrophysicists concerned with the nature of supermassive black holes have been asking themselves a cosmological "chicken and the egg" question: "Which came first, the gargantuan pit of closed spacetime, or the lively panorama of gilded stars and glowing gas that we call a galaxy?"

Prior to a remarkable recent discovery that now seems to have answered this question for the majority of cases, the evidence in favor of black holes appearing first was anchored by the telling observation that the number of quasars peaked 10 billion years ago, early in the universe's existence. The light from galaxies, on the other hand, originated much later – after the cosmos had aged another 2 to 4 billion years. Unfortunately, both measurements are subject to uncertainty, and no one can be sure we are measuring *all* of the light from quasars and galaxies, so this argument is not quite compelling. But

---

[1] As we shall see in Chapter 5, supermassive black holes sometimes gulp down matter at such a high rate that the ensuing compression squirts some of it out before the gas can all be absorbed through the event horizon.

astronomers do see quasars as far out as they can look, and the most distant among them tend to be the most energetic objects in the universe, so at least *some* supermassive black holes must have existed at the very beginning of all things.

Not to be outdone, the proponents for galaxies appearing first point to images such as Fig. 1.5, which show enormous systems of stars and gas assembling from the merger of smaller structures – yet no quasar is visible in its vicinity. Perhaps not every collision feeds a black hole or, what is more likely, at least some galaxies must have formed first.

Astrophysicists are now concentrating their attention on three prominent scenarios for the creation of supermassive black holes in the universe. In every case, growth occurs when matter plummets into the bottomless pit of contorted spacetime, either following the collapse of massive gas clouds, or via the catabolism of smaller black holes in collisions and mergers.

4.1 PRIMORDIAL SEEDS

All of the structure in the universe traces its beginnings to a brief era shortly after the Big Bang. Very few "fossils" that scientists can use to unravel the mystery of the early universe remain from this period; one of the most important is the cosmic microwave background radiation – the afterglow left over from this immeasurably hot, chaotic genesis. In the theory of the Big Bang, most of the hydrogen was dissociated into freely floating protons and electrons, forming a hot sea of charges and currents coupled to electric and magnetic fields. All particles and radiation were in equilibrium with each other, colliding frequently and exchanging energy liberally; matter and radiation behaved as a single fluid.

The rapid expansion that ensued lowered the matter density and temperature, and about one month after the Big Bang, the rate at which light was created and annihilated could no longer keep up with the thinning plasma. The radiation and matter began to fall out of equilibrium with each other, forever imprinting the conditions of that era onto the photons that reach us to this day from all directions in space.[2] The very existence of the cosmic microwave background radiation appears to rule out the steady state model of the universe, for which changes occur without a gradient in time, so that any one step in the evolution of the cosmos would look like any other. This picture is inconsistent with the background radiation since the universe today does not contain the type of environment required to produce it.

The all-pervasive microwave background radiation was discovered serendipitously in 1965, but its story begins many years earlier. In 1934, not long after Hubble discovered the expansion of the universe, Richard Tolman (1881–1948) was a professor of chemistry at CALTECH, though his interest had already started to drift away from physical chemistry, toward cosmology. Early in his career, he had demonstrated that the electron was the charge-carrying particle in metals and even determined its mass, but he was best known later in life for the masterful treatise he published that year – *Relativity, Thermodynamics, and Cosmology* – which has since introduced numerous students to the intricacies of general relativity. The final paragraph of that formidable work[3] is reproduced here, for it illustrates not only his approach to science and philosophy, but perhaps best encapsulates the essence of what cosmology is all about:

> It is appropriate to approach the problems of cosmology with feelings of respect for their importance, of awe for their vastness, and of exultation for the temerity of the human mind in attempting to solve them. They must be treated, however, by the detailed, critical, and dispassionate methods of a scientist.

As the universe evolved, Tolman explained, its temperature would have dropped and its photons would have been redshifted by the cosmological expansion to longer wavelengths. The first reliable

---

[2] An excellent, though somewhat technical, account of the early history of the universe may be found in Linde (1990).

[3] Richard Tolman's book has been reissued by Dover Publications and is again finding a significant readership among those interested in the history of cosmology.

prediction of the radiation temperature was made years later, by George Gamow, Ralph Alpher, and Robert Herman,[4] who were investigating the idea that the chemical elements might have been synthesized by thermonuclear reactions in the primeval fireball. By the present epoch, they estimated, the radiation temperature would have dropped to very low values, as low as 5 degrees above absolute zero.

Thinking back to the period between 1930 and 1965, during which Hubble's cosmology was finding its roots, historians could not be faulted for viewing this development as the quintessential example of how dislocated and serendipitous discoveries congeal into a scientific discipline only with the unhindered perspective and focusing power of hindsight. The cosmic microwave background radiation, it turns out, was actually discovered – though not recognized – well before Gamow, Alpher, and Herman began tinkering with thermonuclear reactions in primeval fireballs. It was rediscovered by accident in 1965, again escaping proper identification, but we will return to this in a moment.

In 1940, World War II had just begun its rage of excesses in Europe when Andrew McKellar, a 30-year-old astronomer working at the Dominion Astrophysical Observatory in British Columbia, Canada, pointed his telescope toward the bright star zeta Ophiuchi. He was trying to prove a conjecture made earlier by Pol Swings and Léon Rosenfeld in Belgium that several unidentified features of the radiation produced by the interstellar medium were probably due to simple diatomic molecules. One can think of these particles as tiny dumb-bells – two side lobes connected with an intermediate spring. Absorbing light and colliding with other matter, these molecules can either vibrate back and forth or swivel about an axis of rotation, in each case giving off the telltale radiative signature of their motion.

Not only did McKellar confirm the existence of these excited molecules in the medium between the stars, but in the process also

---

[4] These individuals published several papers on this subject, some as co-authors, others separately. Two key reports were published by *Nature* in 1948, one by Gamow and the second by Alpher and Herman.

made his most important contribution to science – he discovered an *in situ* thermometer to measure the temperature of space. Each diatomic molecule rotates with an energy commensurate with the radiative heat it absorbs from its environment – the higher the temperature, the faster it spins. By comparing the number of tiny dumb-bells rotating with a variety of energies, McKellar deduced a "rotational" temperature of 2.3 Kelvin for the gas in the interstellar medium.[5] After a flurry of papers in 1941, McKellar seems to have written nothing more on this subject, and passed away in May 1960. He unfortunately never realized the significance of his work, which would not find contextual meaning for at least another five years.

The subsequent development of experimental cosmology owes thanks to the commercialization of space for facilitating the next series of advances – starting with yet another disconnected discovery. Arno Penzias and Robert Wilson had been employed by Bell Labs in Crawford, New Jersey, to use a sensitive microwave horn radiometer as a link to the early Telstar telecommunications satellite. Bell Labs eventually decided to abandon this business in 1963, freeing Penzias and Wilson – radio astronomers by training – to recycle the radiometer for use in more fundamental scientific investigations. Shortly thereafter, while studying the radio emission from the Cassiopeia A supernova remnant, they detected a uniform source of noise that at first seemed to be produced by the apparatus. It took many months of checking and rechecking the equipment and its electronics, including the removal of a bird's nest from the horn, to convince themselves that the signal was actually coming from the sky.

Ironically, a group at Princeton University was just about ready to test its own receivers in a search for the cosmological radiation, and the frustration felt by Penzias and Wilson was alleviated a few months later when their discovery was confirmed by the Princeton researchers. Penzias and Wilson published their result in a very brief

---

[5] This was quite a remarkable achievement, considering that the best value we have now, six decades later, is 2.73 Kelvin. See McKellar (1941).

paper[6] with the unassuming title "A Measurement of Excess Antenna Temperature at 4080 Mc/s." Beautifully complementing this historical measurement, the Princeton group published a companion paper explaining its cosmological significance. In 1978, Arno Penzias and Robert Wilson won the Nobel Prize in physics for their report, believed to be one of the shortest ever to be so honored.

Cosmology had finally merged into the mainstream of science, and the cosmic microwave background radiation could now be probed for the fossilized treasures waiting to be discovered. Astronomers hoped that they would eventually understand why the universe has structure, why galaxies formed, and how supermassive black holes were granted a license to grow. Theorists realized quickly that fluctuations in the temperature of the cosmological radiation would be a reflection of the initial perturbations in density that presumably grew into the mass condensations seen later in the aging universe. Given such enthusiasm, it is not surprising now that the initial estimates of how large these perturbations would be – one part in a hundred – were greatly exaggerated. This level of sensitivity was attained by the improving instrumentation after only a few years, and the background radiation still seemed to be uniform on that scale.

The ensuing quarter of a century saw a rather curious struggle to pin down the temperature anisotropies, in which theorists continually revised their estimates downward to keep ahead of the experimenters' increasingly stringent upper limits. As far as they could tell, astronomers still saw what appeared to be a uniform radiation field. And this caused quite a stir when the limit reached the level of one part in a thousand. With fluctuations this small, said the theorists, density perturbations associated with ordinary matter – the stuff of which we are made – would not have had sufficient time to evolve freely into the nonlinear structures we see today. Only a gravitationally dominant dark-matter component could then account

---

[6] See Penzias and Wilson (1965); the companion paper was published by Dicke et al. (1965) in the same journal.

for the strong condensation of mass into galaxies and supermassive black holes. The thinking behind this was that whereas the cosmic microwave background radiation interacted with ordinary matter, it would retain no imprint at all of the dark matter constituents in the universe. The nonluminous material could therefore be condensed unevenly (sometimes said to be "clumped") all the way back to the Big Bang and we simply wouldn't know it.

Astrophysicists know of at least three reasons why fluctuations in the density of ordinary matter should produce patches in the observed intensity (or temperature) of the cosmic microwave background radiation. The first is simply a change in the intrinsic temperature of the plasma where the radiation is produced. Gas heats up when it gets compressed, and the fractional temperature change equals the fractional density perturbation. Second, clumps of gas that are moving toward or away from us will Doppler-shift the microwave radiation they produce, just as the disk in MCG-6–30–15 (see Section 3.5) is apparently Doppler-shifting the X-rays produced near its event horizon. Third, a clumping of matter in the early universe would have produced variations in local gravity, so the cosmic radiation passing through these peaks and valleys would have undergone myriad intensifications and degradations along the way. (Incidentally, this would apply to dark matter fluctuations as well, but the effect is not as large as the others.)

So how far down in sensitivity did astronomers have to go before they could actually see these fossilized patches in the sky? Finally, in 1992, the Cosmic Background Explorer (COBE) satellite detected the much anticipated fluctuations, bringing some measure of relief to the rapidly growing number of cosmologists.[7] In the all-sky map shown in Fig. 4.1, the temperature of the background radiation is displayed on a scale such that red regions are 0.0002 Kelvin hotter than the cold regions, shown in blue. Discounting the red equator in this image, which represents primarily the microwave radiation produced within our

---

[7] See Smoot and Davidson (1993).

own galaxy, we are witnessing fluctuations that remarkably trace the variations in matter density imprinted in the early universe, shortly after the Big Bang.

Producing this image required a very careful subtraction of other effects associated with the satellite's motion through the cosmos. Its velocity about the Earth, the Earth about the Sun, the Sun about the galaxy, and the galaxy through the universe, also make the cosmic microwave background radiation seem more intense (or hotter) – by about one part in 1000 – in the direction of motion compared to the reverse. The magnitude of this effect, arising from a Doppler-shift, gives astronomers the opportunity of determining the velocity of our local group of galaxies relative to the underlying fabric of space itself. We are apparently drifting with a speed of 600 kilometers per second with respect to the general universal expansion in our neighborhood.

The COBE satellite was equipped with instrumentation that could measure not only the sense of a forward–backward asymmetry, but could also uncover tiny fluctuations on angular scales in the sky that correspond to a distance of about 1 billion light-years. This is still larger than the largest material structures astronomers see in the cosmos, but is nonetheless adequate for them to confirm the notion that the early universe was not perfectly homogeneous. The patches of color we see in Fig. 4.1 represent temperature fluctuations that amount to no more than one part in 100 000 – hardly greater than the accuracy of the measurements. However, the angular variations appear to differ statistically from random noise, and so these represent the first evidence for a departure from exact isotropy that theoretical cosmologists have long predicted to be the seeds of structure in the universe.

On scales much smaller than this, inhomogeneities in the cosmic microwave background radiation would have also been produced by "echoes" of the Big Bang. Matter moving at the speed of sound had sufficient time, before protons and electrons combined to form hydrogen and helium, to move the distance spanned by an angle of about 1 degree of arc on the sky.[8] The ensuing oscillations reverberated across the universe, like the sound waves piercing the air after an explosion. The temperature variations resulting from this process are called "acoustic" fluctuations, and the scale associated with how far a sound wave moves from the beginning of the Big Bang to when hydrogen recombines is known as the "sonic" horizon. At the time of writing, several observational campaigns have reported the preliminary results of their high-resolution mapping of the cosmic microwave background radiation and the results are consistent with models of the early universe.

Incidentally, there is a very important flip side to the fact that astronomers have had an incredibly difficult time seeing fluctuations in the cosmic microwave background radiation. The fact that it exhibits such a high degree of isotropy represents both an aesthetic gratification and a difficulty for the simplest theories. Such homogeneity and isotropy are difficult to explain because of the "light-horizon" constraint. Look up at the sky and imagine being able to sense the microwave radiation coming at you from two opposite sides of the universe. Traveling at the speed of light, the photons are just now arriving at Earth from the distant hot plasma that spawned them. Thus, the matter on one side of the sky could not possibly have had time to communicate with its counterpart on the opposite side – they are beyond each other's light horizon. So how is it possible that all of the hot plasma in the early universe could "know" to have the same temperature with a precision approaching one part in 100 000?

The answer seems to have been provided by the so-called inflationary model of cosmology, in which the early universe underwent

---

[8] The principal groups that have thus far reported their results are the BOOMERanG experiment (Balloon Observations Of Millimetric Extragalactic Radiation and Geomagnetics), the MAXIMA balloon-borne experiment, and the Mobile Anisotropy Telescope (MAT) at Cerro Toco, Chile. These discoveries have been reported by Miller *et al.* (1999), de Bernardis *et al.* (2000), and Hanany *et al.* (2000).

an exponential growth in size.[9] The essential element of this model is that there exists a particle whose nature changes with temperature. As the universe expanded and the temperature dropped, the theory goes, the energy density stored in the vacuum of space did not change because of the presence of an ever increasing number of these entities. If correct, the effect of this particle proliferation on the expansion of the universe was considerable, leading to an exponential growth in size – in essence, an inflation. Eventually, this energy was transformed into the thermal motion of matter, and the universe again became extremely hot, after which its evolution would be described by the standard hot universe theory, with the important refinement that the initial conditions for the subsequent expansion of the hot universe were determined by processes that occurred earlier, during the period of inflation. The inflationary model of cosmology still contains several unresolved issues, and variants are now appearing that address some of the remaining problems. Nonetheless, the nearly perfect isotropy of the cosmic microwave background radiation shows that the entire observable universe had to be causally connected prior to the time at which the radiation decoupled from matter, and this is strong evidence for an earlier period of inflation.

The first billion years of evolution following the Big Bang must have been quite a show, with the various players all jostling for primacy and lasting influence on the structure we see today. The issue of how the fluctuations in density, mirrored by the uneven cosmic microwave background radiation, eventually condensed into supermassive black holes and galaxies is a topic of ongoing research and vigorous debate. It is one of the most important questions in science, dealing with the fundamental contents of the universe, and possibly what produced the Big Bang and what came before it. As we shall see shortly, the evidence now points to a coeval history for these two dominant classes of objects – supermassive black holes and galaxies – though

[9] The foundations for this theory were provided by individuals such as Starobinsky (1980) and Guth (1981). See also Linde (1990), who includes a summary of his early work on this subject in 1979–81.

as we have already noted, at least some of the former must have existed well before anything else. So how did they come about?

An interesting idea being pursued by Stuart Shapiro and his collaborators at the University of Illinois in Urbana/Champaign[10] is that the first supermassive objects formed from the condensation of dark matter alone, without the participation of "ordinary" matter; only later would these seed black holes have imposed their influence on the latter. But this dark matter has to be somewhat peculiar, in the sense that its constituents must be able to exchange heat with each other. As long as this happens, a fraction of its elements evaporate away from the condensation, carrying with them the bulk of the energy, and the rest collapse and create an event horizon. The net result is that the inner core of such a clump forms a black hole, leaving the outer region and the extended halo in equilibrium about the central object. Over time, ordinary matter gathers around it, eventually forming stars, planets, and life – which then begins to make us wonder about the meaning of it all.

The problem with ordinary matter collapsing to form the first supermassive black holes on its own is that initially it was simply not sufficiently clumped, as revealed by the frailty of the fluctuations appearing in Fig. 4.1. Perhaps this material formed the first stars, and then more stars, eventually assembling a swarming cluster of colliding timebombs. Over time, the inner core of such an assembly would have collapsed due to the evaporation of some of its members and the ensuing loss of energy into the extended halo, just as the dark matter did.[11] It only takes a small black hole to start the process of coagulation. Estimates show that, once formed, an object such as this can double its mass every 40 million years so, over the age of the universe, even a modestly appointed dark pit could have grown

[10] Some highly relevant publications dealing with this train of thought include those by Zel'dovich and Podurets (1965), Shapiro and Teukolsky (1992), and Balberg and Shapiro (2002). See also Umemura, Loeb, and Turner (1993) and Eisenstein and Loeb (1995).

[11] The earliest proposal for such a process seems to have been made by Lynden-Bell and Wood (1968). See also Quinlan and Shapiro (1990) and Haehnelt and Rees (1993).

into a gargantuan billion-solar-mass object patrolling the cosmic frontier.

Some surprisingly compelling evidence that this scenario must have operated in the early universe has been uncovered recently by the Hubble Space Telescope (see Fig. 4.2). Globular clusters contain the oldest known stars. Assembled over 12 billion years ago, many of these objects are truly the fossilized record of the earliest era of galaxy formation. Probing the stellar dynamics of G1, a globular cluster orbiting the Andromeda galaxy, the Hubble Space Telescope identified a 20 000-solar-mass black hole lurking in its center. This object did not grow into megalithic proportions due to the paucity of other matter in its vicinity. Many of its brethren, however, possibly endowed with a more favorable disposition, could very well have grown exponentially before the universe thinned out from expansion.[12]

In either case, the evidence no longer points to these mechanisms as being solely responsible for the creation of the millions of supermassive black holes sprinkled throughout the universe – indeed, most of them could not have formed in isolation, but neither could their host galaxies.

4.2 GALAXY TYPES

Galaxies change with time. Some scar with age; others collide, merge, and recycle into bigger, bolder arrangements. Many of the earliest crucibles of star formation may have looked like the globular cluster G1 in Fig. 4.2. But the sideways motion of gas condensing out of the primordial fluctuations (see Fig. 4.1) would not have been easily lost, so rotation was a defining characteristic of many galactic profiles in the pre-adolescent universe. Probably not recognized at first, the general sequence comprising these cosmic communities of huddling stars constitutes a "timeline" of sorts, whose viability grows as an increasing number of supermassive black holes are linked synergistically to their hosts.

---

[12] The discovery of an intermediate-size black hole at the center of the globular cluster known as M15 was reported by van der Marel (1999). The discovery of a black hole at the nucleus of G1 was announced in 2002 by Karl Gebhardt and his collaborators.

Since galaxies constitute the bulk of the matter that emits visible light, their existence was noted long before the advent of powerful telescopes and photography that would later record and identify them. Astronomers had observed and cataloged certain nebulous patches and relatively bright sources of light as far back as the beginning of the eighteenth century. Indeed, the French comet hunter Charles Messier carefully described these objects and assembled them in a catalog now known as the *Messier Catalogue,* listing over a hundred sources that we recognize today as galaxies, star clusters, and gaseous nebulae. The Milky Way's sister galaxy in Andromeda is the 31st entry on that list, hence its designation as M31. Since then, several other more comprehensive catalogs have been compiled, though as we noted earlier in this book, it wasn't until the mid-1920s that Hubble – and subsequently other astronomers – recognized these patches of light as constituting entirely separate aggregates of stars external to our own galaxy.

It was Hubble, in fact, who built a classification scheme for galaxies, arranging them in an orderly progression that we now call the Hubble sequence. Basing his order on 600 well-defined bright objects, he organized them starting with essentially spherical and elliptical configurations, through lens-shaped systems to very flat spiral galaxies. The irregularly shaped ones form a separate class.

Spherical and elliptical galaxies are distinguished from the three other main classes (lenticular, spiral, and irregular) by their very smooth and symmetrical texture and no evidence of an internal structure. They possess a glowing center and their brightness fades towards the edges. They show no evidence for a disk, or plane, and certainly no spiral structure like our own galaxy. Centaurus A in Fig. 1.6 is an almost perfect elliptical structure, though it clearly retains dark dust lanes presumably left over from an earlier collision and merger with a smaller spiral galaxy. The constituent stars in these systems tend to be very old, and very few new stars are currently forming in them. Astronomers interpret this to mean that, like globular clusters, elliptical galaxies harbor the first stars to emerge in the nascent universe. But ellipticals were not born that way; they are by-products of hierarchical collisions and mergers of smaller – presumably disklike – galaxies. As we shall see shortly, the spheroidal stars in the central bulge of our own galaxy form a collection that is very much like that of an elliptical galaxy, though significantly smaller in number.

Lenticular and spiral galaxies (two separate classes in Hubble's sequence) contain a disk of stars and generally also a central spheroidal component. Spirals are also distinguished by two or more arms emerging from opposite sides of a nucleus, winding through the disk and tapering off toward the galaxy's extremity. Our own galaxy, and that in Andromeda, are among the most conspicuous members of the spiral class. Another very beautiful example in this category is the Sombrero galaxy (M104) in Fig. 4.3, captured recently with the Very Large Telescope ANTU in Paranal, Chile. It has been determined that the Sombrero galaxy's nucleus emits a very potent flux of X-rays, and coupled with the unusually high stellar velocities measured in its central region, the X-ray profile leads astronomers to speculate that a one-billion-solar-mass black hole resides there. This galaxy is also notable for its dominant nuclear bulge, composed primarily of mature stars; its nearly edge-on disk is composed of stars, gas, and intricately structured dust. Quite generally, the youngest, brightest stars in disk galaxies (either lenticular or spiral) tend to emerge in the planar region and, if present, primarily within the spiral arms. In contrast, the central hub resembles elliptical galaxies in content and stellar demography.

In recent years, astronomers have discovered over 10 000 peculiar galaxies that do not fit into any of Hubble's classes. Many of these are obviously undergoing collisions (for example, the Antennae galaxies in Fig. 1.5) or are survivors of recent catastrophic events. Some are members of bound systems connected to each other by luminous bridges of stars and glowing interstellar matter. Another small – though critically essential – subgroup of the peculiar category are galaxies with unusually active cores, displaying a broad range of phenomena associated with very violent or energetic events in their nuclei.

Functioning more as cosmological fossil hunters than scientists testing current phenomena, astronomers struggle to piece together the history of how these classes of galaxies came to be. How are they related to supermassive black holes? How did they form? And what role did quasars play in galaxy evolution? A decade ago, any attempt to corral supermassive black holes into an analogous classification scheme would have produced a rather simple answer – they're an oddity, an exception to an otherwise harmonious, galaxy-dominated universe. Times have changed, and as a prelude to our discussion of their newly recognized symbiotic co-evolution with galaxies, we shall next consider how prevalent these objects really are.

4.3 THE SUPERMASSIVE BLACK HOLE CENSUS

Each of the supermassive black holes we have looked at in this book is special for one reason or another, either because it was the first quasar ever discovered (3C 273; see Fig. 1.2), or because it is the nearest radio galaxy (Centaurus A; see Fig. 1.6), or because it is the active galactic nucleus with arguably the most spectacular pair of jets and radio lobes (Cygnus A; see Fig. 1.8). But there are billions of galaxies extending, as far as we can tell, to the edge of the visible universe, where the evidence for the existence of supermassive black holes first appears. How many of these peculiar objects are there, we wonder?

Many astronomers suspect that almost every large normal galaxy harbors a supermassive black hole at its center, a hypothesis for which supporting evidence is gradually growing. In a recently completed survey of 100 nearby galaxies using the Very Large Array radio telescope in New Mexico,[13] followed by closer scrutiny with the Very Long Baseline Interferometry array, at least 30 percent of this sample showed tiny, compact central radio sources bearing the unique signature of the quasar phenomenon. Their faint glimmer appears to be the relic signature of headier days long past, when some of

---

[13] This work was carried out by A. Wilson from the University of Maryland and his collaborators.

these objects may have been the powerful beacons shining from the universe's epoch of structure formation.

In another notable development, NASA's Chandra X-ray telescope appears to have settled a long-standing puzzle, dating back to the early 1960s. The universe, it turns out, is aglow with the faint murmur of diffuse X-ray light, which fills the entire sky. Using a deep exposure of a selected piece of the heavens,[14] Chandra was able to resolve at least 80 percent of the X-ray glow into myriads of individual point sources, suggesting an extrapolated total number of about 70 million across the entire sky. In follow-up studies of some of these objects, using telescopes to sense their radiation at other wavelengths, the researchers who carried out these observations concluded that some are relatively normal galaxies with dust-veiled X-ray-emitting nuclei – the signature of a central black hole.

Others are very distant quasars, too faint to shine brightly like those in Fig. 1.2. And the rest are either unknown or not related to supermassive black holes. The picture that emerges from surveys such as these is that the universe is awash with pockets of isolated spacetime. A subset of them formed very early, probably even before the primordial age when galaxies were first coagulating from the fragmentary gas clouds condensing out of the Big Bang. Some of them may be true fossils from the cosmological "Dark Age," the period extending over several hundred million years between the cooling of the Big Bang and the epoch of star formation. However, all of them now seem to reside in the nuclei of galaxies – it is telling that no isolated supermassive black hole has ever been found.

Still, the more conservative among us would argue that these distant points of light are highly suggestive of the presence of supermassive black holes, but that this conclusion is only tentative until we actually see them, or deduce their presence from more compelling evidence, as we did in Chapter 2 for Sagittarius A* at the galactic center and for NGC 4258 with its maser-emitting central disk. Spearheading

[14] See Mushotzky et al. (2000).

this approach, John Kormendy at the University of Texas and Doug
Richstone at the University of Michigan, and their collaborators, have
set about the task of meticulously assembling the clues required to
complete the detective work for as many of these objects as is currently feasible. At the time of writing, direct measurements of supermassive black holes have been made in over 38 galaxies, based on the
large rotation and random velocities of stars and gas near their centers.
These objects are all relatively nearby because these direct methods
do not work unless we can see the individual stars in motion about
the central source of gravity.

And now the first curious piece of the puzzle emerges – none of
the supermassive black holes have been found in galaxies that lack a
central bulge (see Fig. 4.4). As we have seen, galaxies come in two basic
types (excluding irregulars for now) – those that contain flat, spinning disks, and others that have more nearly spherical bulges that rotate only a little and are comprised mostly of randomly-moving, older
stars. Many galaxies, such as the Milky Way and Andromeda, consist
of a disk *and* a bulge in the middle. So far, astronomers have found
a supermassive black hole in every galaxy observed that contains a
bulge component, but none in those that only possess a disk and no
central hub. The former may have undergone one or more mergers in
their past, whereas the latter are presumably pristine condensations
of swirling matter untouched since primeval times. Thus, the first
clue we gather from these surveys is that a collision, like that seen
in Fig. 1.5, may have been required to create a central supermassive
object, a process in which the highly-ordered cartwheel structure of
the "bulgeless" galaxies is at least partially disrupted.

Uncovered only recently,[15] a second clue now appears to have
clinched the case for a coeval growth of most galaxies with their supermassive black holes. Rather than being a destructive influence
on the universe, the latter appear to have been essential to the very

---

[15] These results were produced by two groups working independently, and reported by
Ferrarese and Merritt (2000), Merritt and Ferrarese (2001), and Gebhardt *et al.* (2000).

creation of the structure within which they live. Researchers at the Rutgers State University of New Jersey and at the University of Texas in Austin have shown quite convincingly that the mass of the central black hole can be predicted with remarkable accuracy simply by knowing the velocity of stars orbiting within the spheroidal component of the host galaxy. Pinpointing the trajectory of a single object in that maelstrom of light and motion is difficult and impracticable. Instead, astronomers can measure the accumulated light from a limited region of the galaxy's central hub and extract, from the collective Doppler shift they infer, an average speed of the stars as a group. What they have found is that the ratio of the black hole's mass to this average speed is constant across the whole sample of galaxies they surveyed.

This finding is one of the most surprising and remarkable correlations ever discovered in the study of how the universe acquired its structure. Taken at face value, one would think there ought not to be a link between the supermassive black hole and the motion of stars in the outer reaches of the host galaxy's bulge, where the gravitational influence due to the former is completely nonexistent. Supermassive black holes, it seems, "know" about the motion of stars that are too distant to feel their gravity directly. Obviating the simple naive picture of a ponderous dark pit rampaging aimlessly through the primeval soup, this tight connection instead compels us to postulate an entangled history between a central black hole and the beehive stellar activity in its halo. Although they may not be causally bonded today, they must have had an overlapping genesis in the past.

It turns out that about 0.1 percent of a galaxy's luminous mass is associated with its central black hole. Viewed as an ensemble, these objects have a density comparable to that expected on the basis of the observed number of quasars, whose terminal endpoints are now viewed as being the relatively dormant behemoths lurking in the nuclei of otherwise "normal" galaxies. With these facts in hand, and satisfied that most supermassive black holes and their host galaxies grew symbiotically, astrophysicists hypothesize that, once created, a primordial condensation of matter continues to grow with a direct feedback on its surroundings. This may happen either because the quasar heats up its environment and controls the rate at which additional matter can fall in from its cosmic neighborhood,[16] or because mergers between galaxies affect the growth of colliding black holes in the same way that they determine the energy (and therefore the average speed) of the surrounding stellar distribution.

Many astronomers now believe that in the majority of cases, the merger of galaxies – past or present – were ultimately behind the hierarchical construction of elaborate galaxies with elliptical, or disk-plus-bulge, profiles.[17] In this paradigm, the small density inhomogeneities that formed in the dark matter shortly after the Big Bang grew as the universe expanded, eventually collapsing into relatively small objects like the G1 globular cluster now in orbit around Andromeda (Fig. 4.2). Larger galaxies grew with wave upon wave of collisions and mergers, a process contributing significantly to the variety of shapes encompassed by the Hubble sequence. As we shall see shortly, detailed numerical simulations of the merger of two spiral galaxies of about equal mass produce remnant galaxies structurally very similar to galactic bulges. Meanwhile, the twisting of gravity at the heart of this encounter drives most of the gas into the middle, where it can form new stars and feed a central black hole, or a pair of black holes. The tight correlation between the black-hole mass and the average speed of the stars in its halo appears to be a direct consequence of this cosmic cascade.

[16] In support of the idea that massive black holes may have formed prior to the epoch of galaxy definition, Silk and Rees (1998) suggest further that protogalactic star formation would have been influenced significantly by the quasar's extensive energy outflows. The ensuing feedback on the galaxy's spheroidal component could be the reason we now see such a tight correlation between the mass of the central object and the stellar velocities much farther out.

[17] The consensus among modelers of galaxy formation seems to be that most large galaxies have experienced at least one major merger during their lifetime. See "How are Galaxies Made?" *Physics World*, May 1999, 25–30.

4.4 GALAXY COLLISIONS

The quasar QSO 1229 + 204 is a good fraction of the way across the observable universe, yet it is so bright that only the high resolving power of the Hubble Space Telescope could separate out its powerful radiation from the much fainter glow of the host galaxy (see Fig. 1.3). Remarkably, QSO 1229 + 204 is not only at the core of an unusual spiral galaxy with a bar across its middle, but it is actually in the process of colliding with another, smaller, galaxy. Almost certainly, gas churned by this collision is being funneled into the active, turbulent core, where it fuels a supermassive black hole, causing it to shine so brightly.

We ourselves are active participants in the galaxy collision game, as we shall soon learn when we next consider the future consequences of Andromeda's aggressive acceleration toward the Milky Way. Until very recently, however, the end product of such a galactic catastrophe was still uncertain, given that astronomers had yet to identify a merger remnant whose origin was clearly two colliding nuclei. That changed abruptly at the end of 2002, when for the first time, scientists obtained proof that two supermassive black holes exist together in the same galaxy. Focusing on the core of NGC 6240, a butterfly-shaped galaxy believed to be the product of a collision between two smaller galaxies some 30 million years ago, NASA's Chandra X-ray Observatory produced an image of these two objects orbiting each other with a separation of only 3000 light-years (see Fig. 4.5). Several hundred million years from now, their orbit will have decayed to create an even larger black hole in a spectacular cosmic flash that will unleash an intense burst of radiation and gravitational waves.

This breakthrough resulted from Chandra's ability to clearly distinguish closely separated objects at great distances, and to therefore measure the details of the X-radiation from each nucleus. The black-hole nature of both sources was revealed by the excess number of high-energy photons each of them produces within the surrounding hot swirling gas, and by the radiation emitted by free-floating iron atoms, as was the case in the spinning black hole MCG-6–30–15 (see Chapter 3).

Like the host of the quasar QSO 1229 + 204, NGC 6240 is a massive galaxy undergoing star formation in its nucleus at an exceptionally rapid rate due to the recent collision. The large quantity of dust and gas driven into the middle during the encounter makes it difficult to peer deeply into its central region with optical telescopes, but the X-rays can easily penetrate through the obscuring shroud, producing this magnificent view of one of the most dramatic phenomena in the post-Big-Bang universe. Massive black holes evidently can grow through the merger of smaller collapsed objects, perhaps even producing a burst of gravitational waves detectable with future space-borne instruments.[18]

Led by their director, Bernard Schutz, scientists from the Max Planck Institute for Gravitational Physics (also known as the Albert Einstein Institute) in Golm near Potsdam have been simulating grazing collisions of two massive black holes on supercomputers, hoping to identify the cosmic fingerprint carried by the gravitational waves emitted during such an encounter. In a galactic merger, the supermassive objects sink rapidly to the center of the fused system because of the "friction" they experience moving through the surrounding cluster of stars. Losing energy along its trajectory, each black hole drifts closer and closer to its partner, eventually forming a very tight binary in the middle.

Looking past the current generation of instruments designed to detect primarily the radiation produced by these sources, the most exciting technological frontier to be assailed next is the detection of gravity waves. Just as two buoys rotating about each other on placid water produce outwardly seeking undulations, the two orbiting black

---

[18] The discovery of the twin supermassive black holes in the nucleus of NGC 6240 was based on observations carried out with the Advanced CCD Imaging Spectrometer on the Chandra satellite by Susan Komossa, Gunter Hasinger, V. Burwitz, and P. Predehl at the Max-Planck-Institut für Extraterrestrische Physik in Germany, J. Kaastra at the Space Research Organization in the Netherlands, and Y. Ikebe at the University of Maryland in Baltimore.

holes distort the surrounding spacetime by periodically raising and lowering the intensity of gravity. These disturbances probe away from the system at light speed, spreading throughout the universe and producing ripples in the fabric of space. Locally, these would appear as minute changes in the distance between any two points.

The simulations carried out by the physicists at the Albert Einstein Institute, and their colleagues at Washington University in St Louis and the Konrad-Zuse-Zentrum in Berlin, show that during the last few moments the black holes spiral inward, emitting weak periodic gravitational signals. The event horizons of the two objects stretch and, in a matter of milliseconds, coalesce like raindrops. The strength of the signal, and its frequency, increase rapidly, before disappearing as the oscillations of the unified event horizon damp down and the final black hole becomes quiet.

These calculations reveal that as much as 1 percent of the total energy available during the encounter can be converted into gravitational waves. This is thousands of times more than the entire energy liberated by our Sun during the past 5 billion years. But many of the biggest crashes in the universe occur so far away that the signals reaching Earth are extremely weak. The effect here would be to induce oscillations that would jerk masses spaced 1.1 kilometers apart by one thousandth of the diameter of a proton.

Gravitational waves have never been detected directly, though the idea that the effect of gravity itself should travel at the speed of light gained considerable experimental verification in the latter part of the twentieth century. The binary pulsar PSR 1913 + 16 was seen to have a decaying orbit, attributed to the loss of energy carried off by the escaping gravitational waves. The rate at which this energy is lost depends on the finite speed of propagation, and the orbital changes can therefore equivalently be viewed as a measure of this velocity. The fact that this damping occurs at all is a strong indication that something (i.e., a gravitational wave train) is leaving the system and that it must be doing so with a finite speed – otherwise, the whole collapse would occur instantaneously. The actual measurement confirms that the speed of gravity is equal to the speed of light to within an accuracy of 1 percent.[19]

An actual detection of gravitational waves, when compared with the predictions physicists are now making, would provide direct information on the masses of the coalescing black holes, their spins, orientations, and perhaps even their separation. The prospect for making such fundamental measurements is one of the primary motivations behind the Laser Interferometer Space Antenna (LISA), a space-based gravitational-wave telescope currently under development for a launch after 2010. Ground-based gravitational-wave detectors already exist, but they are not large enough to detect the long-wavelength gravitational undulations produced by binary supermassive black holes.

Consisting of three spacecraft flying in a triangular formation 5 million kilometers apart, LISA would detect a passing gravitational wave by sensing the stretching and squeezing of the space between each pair of components. These shifts are very tiny indeed – no more than a fraction of an atom across – but should be detectable with laser interferometers. LISA's sensitivity will be sufficient to detect binary black hole mergers all the way to the edge of the visible universe. So, knowing how frequently galaxies collide and merge throughout the cosmos, astronomers estimate that these spectacular events should make the instrument flutter about once every couple of years.

4.5 COLLISION OF ANDROMEDA WITH THE MILKY WAY

Assuming our species survives over the next 5 billion years, our descendants should be able to witness one of these cacophonous catastrophes firsthand, as the Milky Way collides and then merges with the galaxy in Andromeda, the Milky Way's closest major galactic neighbor (see Fig. 4.6). The remnant, bigger than either of the two participants, will look nothing like them. Instead, our descendants are likely to live within an elliptical structure, comprised of elderly stars that will have survived the transition and a changing of neighborhoods. But

---

[19] For a technical reference on this subject, see Damour (1987).

this will not be the first such encounter that these two galaxies will have experienced. In 1991, as the Planetary Camera then on board the Hubble Space Telescope focused on the nucleus of Andromeda, astronomers were presented with an unexpected sight – the appearance of not one, but two spots, separated by a mere 30 light-years. A flurry of subsequent activity with ground-based telescopes confirmed that two nuclei do indeed exist in this spiral galaxy, and that they are orbiting each other. Speculation has it that one nucleus is a supermassive black hole, whereas the second is the remnant of a smaller galaxy disrupted and eaten long ago by Andromeda.

The Milky Way and Andromeda are approaching each other with a speed of about 480 000 kilometers per hour. Whether we are in store for a head-on collision, or more of a side swipe that will prolong the agonizing encounter prior to the final merger, will depend on Andromeda's precise tangential motion across the sky, which astronomers continue to determine ever more precisely as better, more powerful telescopes are built. The collision itself will take several billion years to fully run its course, so it would be difficult for any individual civilization at that time to fully appreciate the grand scale – in time and space – of the encounter.

Computer simulations, however, do provide a tantalizing view into how the two galaxies will unravel and ultimately form a giant, spheroidal aggregate of stars – resembling an elliptical galaxy in the Hubble sequence. At the present moment, the Andromeda galaxy appears simply as a spindle-shaped smudge of light in the northern autumn sky. At a distance of 2.2 million light-years – roughly 20 times the diameter of our Milky Way galaxy – it is four times the width of the full moon. As the two galaxies approach each other, our descendants will see Andromeda growing ever larger, casting an eerie slither of glowing light across the heavens.

By the time the two galaxies intersect, the familiar Milky Way arch across the sky will be joined by a second arch, and this crossing pattern should last about 100 million years before the two galaxies engage in their initial retreat. During this time, large clumps of cold

Figure 4.2 In the center of this image, captured in 1994 by the Hubble Space Telescope, resides the globular cluster called G1, a large, bright ball of light consisting of at least 900 000 ancient stars. G1 orbits about 170 000 light-years from the nucleus of the Andromeda galaxy (M31), the nearest major spiral galaxy to our Milky Way, and is situated approximately 2.2 million light-years from Earth. By studying the brightness and colors of its fainter helium-burning members, astronomers can deduce that this structure must have formed over 12 billion years ago, shortly after the beginning of the universe. These stars therefore represent a fossil record of the earliest era of galaxy formation. G1 has a total mass of 10 million Suns, making it one of the most massive globular clusters known. The Hubble Space Telescope recently found that the speeds and spatial distribution of stars in the core of this cluster are consistent with the presence of an intermediate-mass black hole lurking there with a mass of 20,000 Suns. The color picture was assembled from separate images taken in visible and near-infrared light. (Image courtesy of R. M. Rich, K. Mighell, and J. D. Neill of Columbia University, W. Freedman of Carnegie Observatories, and NASA)

Figure 4.3 This spectacular image of the Sombrero galaxy (Messier 104) was obtained with the Very Large Telescope ANTU (which means ``the Sun'' in the Mapuche language) on January 30, 2000, at Paranal Observatory in Chile. It is a composite of three exposures in different wavebands. This is an example of a spiral galaxy that contains both a big bright core, and well defined spiral arms, seen here almost edge on. It also has an unusually pronounced bulge with an extended and richly populated system of stellar clusters. Located in the constellation Virgo (the virgin), its distance from Earth is about 50 million light-years. Compare this with the Centaurus A galaxy (shown in Fig. 1.6), which is closer to being elliptical, with very little evidence of a disk component. The dark dust lanes are presumably left over from its earlier collision and merger with a smaller spiral aggregate of stars. (Image courtesy of Peter Barthel et al. at the Kapteyn Institute and the European Southern Observatory)

Figure 4.4 This image of the spiral galaxy NGC 2613 was captured on February 26, 2002, by the Very Large Telescope MELIPAL (which means ``Southern Cross'' in the Mapuche language), in Paranal, Chile. A search for a supermassive black hole in its core produced a null result (Bower *et al.* 1993). By now the preponderance of evidence suggests that highly flattened disk galaxies lacking a significant central hub, or spheroidal component, also lack a supermassive object in the nucleus. On the other hand, every galaxy that does possess a central bulge (like the Sombrero; see Fig. 4.3) also harbors a supermassive black hole. (Image courtesy of O. Caputi, M. Scodeggio, G. Sciarretta, O. Le Fevre, S. Brau-Nogue, C. Lucuix, B. Garilli, M. Kissler-Patig, X. Reyes, M. Saisse, L. Arnold, and G. Manciniand of the VIMOS collaboration, and the European Southern Observatory)

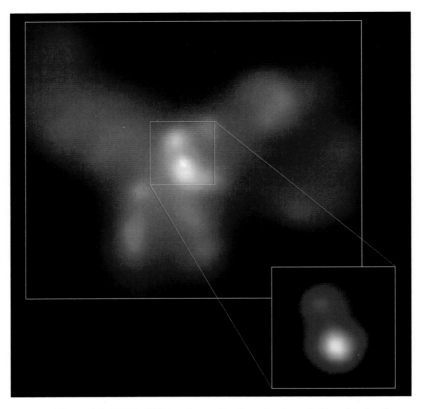

Figure 4.5 NGC 6240 is a butterfly-shaped galaxy, believed to be the
product of a collision between two smaller galaxies some 30 million
years ago. It is a ``starburst'' galaxy in which stars are forming, evolving, and exploding at an exceptionally high rate due to the turmoil
created by the merger. This Chandra X-ray image (34 000 light-years
across) shows the heat generated by the merger activity, which created
the extensive multimillion degree Celsius gas. We see here, for the
first time, direct evidence that the nucleus of such a structure contains
not one, but two active supermassive black holes, based on an excess
of high-energy photons from the gas swirling around each of the bright
objects, and X-rays from fluorescing iron atoms in the nearby gas.
These two black holes are expected to drift toward each other across
their 3000-light-year separation and merge into an even more massive
object several hundred million years hence. (Image courtesy of
S. Komossa, G. Hasinger, and J. Centrella, the Max-Planck-Institut für
Extraterrestrische Physik and NASA)

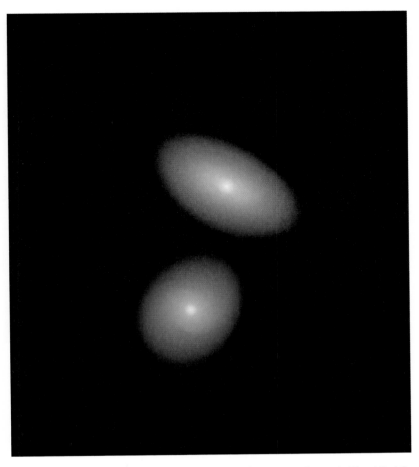

Figure 4.6 This figure, together with the images shown in Fig. 4.8, 4.9, 4.10, and 4.11, are snapshots taken during a computer simulation designed to demonstrate how the collision between the Milky Way and Andromeda might evolve. The calculation was carried out on Blue Horizon, a 1152 processor IBM SP3 at the San Diego Supercomputing Center. Each spiral galaxy is represented by about 40 million stars and is surrounded by a 10-million particle dark-matter halo. The physical size represented by this image is approximately 500 000 light-years across. The Milky Way is here shown face-on approaching from the bottom, whereas Andromeda is tilted and approaching from above. In this first image, the two galaxies are accelerating toward each other under the influence of their mutual gravity. Their shape will not begin to change significantly until they collide. (Image courtesy of J. Dubinski at the University of Toronto)

Figure 4.7 As the collision proceeds, the beautiful spiral disks that for billions of years will have marked the planes of the Milky Way and Andromeda galaxies, will begin to unravel in response to the gravitational pull of the two gargantuan, though now hapless, entities. As Andromeda swings by us, the surrounding sky will light up with brilliant new stars and clusters born in the tattered lanes of gas and dust. This image, taken with the Hubble Space Telescope, shows the interacting pair of galaxies NGC 2207 (the larger, more massive object on the left) and IC 2163 (the smaller one on the right), located some 114 million light-years from Earth in the constellation Canis Major. Alien eyes looking back toward the Milky Way and Andromeda from some distant galaxy might see something like this some 40 million years after the interaction in Fig. 4.6 will have started. By this time, IC 2163's stars have begun to surf outward to the right on a tidal tail created by NGC 2207's strong gravity. (Image courtesy of D. M. Elmegreen at Vassar College, B. G. Elmegreen at the IBM Research Division, NASA, and the Hubble Heritage Team based at the Space Telescope Space Institute and AURA)

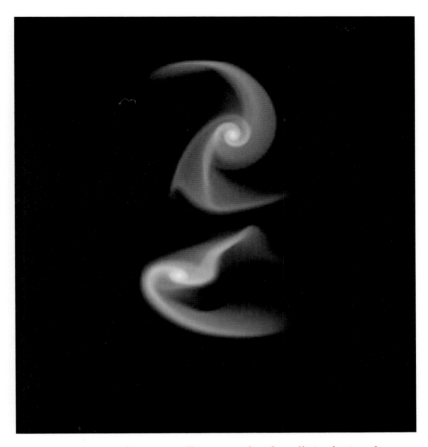

Figure 4.8 About 270 million years after the collision begins, the galaxies swing past each other for the first time and the gravitational interaction excites several open, spiral arms. The Milky Way (top of image) and Andromeda begin to disrupt. The tidal forces of gravity will by then have created long plumes of material, resembling the tails seen in the Antennae galaxies (see Fig. 1.5). (Image courtesy of J. Dubinski at the University of Toronto)

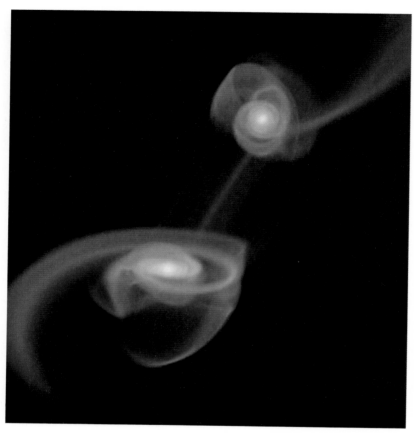

Figure 4.9 In this image, representing the distribution of stars roughly 720 million years after the collision begins, the galaxies fall back together again for a second encounter that further disrupts them, leading to a rapid merger of the material in the central regions. (Image courtesy of J. Dubinski at the University of Toronto)

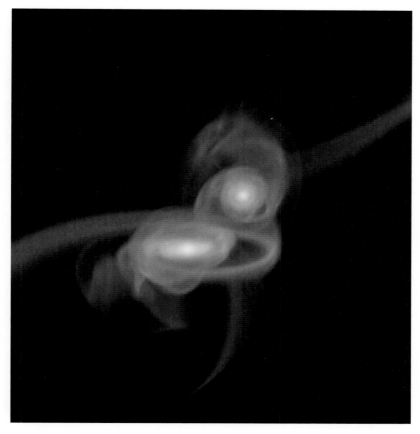

Figure 4.10 By about 1 billion years after the beginning of the collision, the two galaxies engage in a series of convulsions that continue to disrupt their spiral patterns. A portion of each galaxy is swept out into intergalactic space by the expanding tidal tails. (Image courtesy of J. Dubinski at the University of Toronto)

Figure 4.11 Finally, 1.2 billion years after the Milky Way and Andromeda will have begun their gravitational dance, the central region will merge into a single remnant aggregate of stars. In the aftermath, an even more massive black hole than those now lurking in the nuclei of these two galaxies is likely to emerge. Feeding voraciously on the turbulent gases, this supermassive object will ignite to form a quasar. But we would have to wait about 5 billion years to see all of this unfold. Whether our solar system remains as a member of this new galaxy, or is flung out into essentially empty intergalactic space, will depend on where the Sun is located on its orbital trajectory at the time of the encounter. Estimates based on computer simulations such as this indicate that these two outcomes are equally likely. (Image courtesy of J. Dubinski at the University of Toronto)

3 Arc Minutes

Figure 4.12 Chandra studied the nucleus of NGC 253, a relatively nearby galaxy some 10 million light-years from Earth. At this distance, the 3-arcmin segment shown in the bottom-left-hand corner corresponds to a physical dimension of 9000 light-years. Although this object had not shown any prior evidence of a centralized active X-ray region, Chandra uncovered over 10 ultraluminous sources there, three of which are located well within 3000 light-years of the galaxy's core (see the magnified view in Fig. 4.13). If these are middleweight black holes, they may be settling toward the center of the galaxy, creating a full-blown supermassive black-hole system. (X-ray image courtesy of K. Weaver of NASA Goddard Space Flight Center, and D. Strickland and T. Heckman at Johns Hopkins University; black-and-white optical image courtesy of Space Telescope Science Institute and the Digitized Sky Survey)

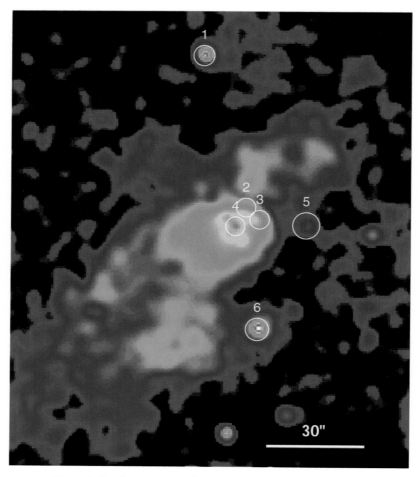

Figure 4.13 This is a magnified view of the X-ray image shown in Fig. 4.12. The 30-arcsec segment at lower right corresponds to a physical size of about 1500 light-years. The ultraluminous sources discussed in the text show up clearly as bright points, numbered from 1 to 6 in this field. The central three objects, believed to be settling gravitationally toward the middle of the galaxy, are numbered 2, 3, and 4. Moving through a dense region of new, massive stars, the three middleweights lose energy due to friction and gradually spiral down into the middle. (Image courtesy of K. Weaver of NASA Goddard Space Flight Center and D. Strickland and T. Heckman at Johns Hopkins University)

Figure 5.3 This is a composite image showing the radio sky above a nighttime optical photograph of the now defunct 300-foot radio telescope in Green Bank, West Virginia, that made the 4.85 GHz radio map. The colors were chosen to indicate increasing radio brightness with lighter shades. This is how the sky would appear to someone with 300-foot-diameter eyes sensitive to low-energy radiation at this frequency. Although many of the sources evident in the radio sky are present also in a parallel image made with optical light, the two views generally contain different classes of object. For example, the brighter radio sources extending from lower left to upper right in this figure lie in the plane of the Milky Way. Some are clouds of ionized hydrogen gas, while the circular or ring-shaped features are remnants of supernova explosions. Another significant difference is that, whereas the stars we see at night are relatively nearby (most within 1000 light-years or so), almost all of the radio sources on this map are luminous radio galaxies or quasars, and their average distance is over 5 billion light-years away. (Image courtesy of J. J. Condon, J. J. Broderick, G. A. Seielstad, NRAO, AUI, and NSF)

Figure 5.4 In Socorro, New Mexico, 27 radio antennas, each with a span of 82 feet and a weight of 230 tons, are arranged in a Y-shaped array whose size can be varied from hundreds of meters to tens of kilometers. The array functions as an interferometer, meaning that the signals from all the antennas are combined electronically, so that the array functions as a single, giant dish. The Very Large Array (VLA) has been in operation since 1980, and is available to astronomers from all over the world. (Image courtesy of NRAO, AUI, and NSF)

Figure 5.5 The main limitation to the resolution attainable in radio imaging is the size of the antenna, or in the case of an interferometric array, the separation of its components. The best radio astronomers can do on Earth is to spread the individual radio dishes across the face of the planet, producing a global baseline. The Very Large Baseline Array (VLBA) is comprised of ten antennas (each like the components of the VLA in Fig. 5.4) spread out across North America from Hawaii to the Virgin Islands. The ensemble is controlled from an operations center in Socorro, New Mexico, and, like the VLA, the individual antenna signals are merged electronically to produce an overall effect comparable to what one would get with an 8000-kilometer dish. The VLBA has been in operation since 1993. (Image courtesy of NRAO, AUI, and NSF)

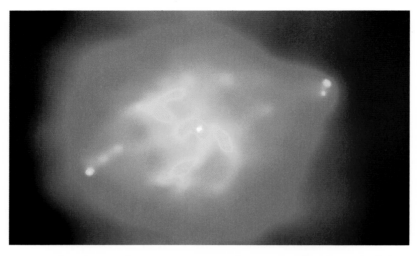

Figure 5.6 With instruments attuned to X-rays rather than radio light, Chandra's spectacular view of the Cygnus A region picks up different physical effects than those sensed by the VLA (see Fig. 1.8). X-rays in this source are produced primarily when relativistic particles pumped out by the central black hole cavitate the thin, hot gas pervading the medium between galaxies. This situation is very much like a balloon being inflated by gas expanding from its middle. Not surprisingly, the brightest features are those at the two termination points where the jets are impacting into the dredged up gas forming the lobes we saw in the radio image. Bright bands around the equator of the enormous oval-shaped bubble are also visible, providing some evidence for the presence of material swirling toward the central black hole. (Photograph courtesy of A. S. Wilson *et al.* at the University of Maryland, and NASA)

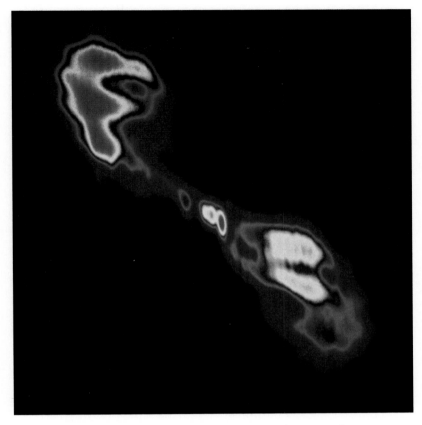

Figure 5.7 The active galaxy Centaurus A, whose optical appearance we first glimpsed in Fig. 1.6, is a frequent target for observations across the spectrum, from radio rays to X-rays and gamma rays. As the archetype for its class, it is one of the most extensively studied objects in the southern sky. The image shown here, taken with the VLA at a wavelength of 2 centimeters, shows a region approximately twice as big as that of Fig. 1.6, though with the same orientation. The core of the jet structure, which coincides with the center of the elliptical galaxy, is the smallest known extragalactic radio source, only 10 light-days across. The jets manifest themselves at radio wavelengths with radiation emitted by high-energy particles, accelerated away from the nucleus by a supermassive black hole with the mass of 200 million Suns . These relativistic particles also produce high-energy radiation that delineates the jets at the other end of the spectrum (see Fig. 5.8). (Image courtesy of J. O. Burns, E. J. Schreier, E. D. Feigelson, NRAO, AUI, and NSF)

Figure 5.8 Chandra's X-ray image of Centaurus A reveals the bright central source — the active galactic nucleus harboring the 200-million-solar-mass black hole — and over 200 other point-like X-ray emitters dispersed throughout the galaxy (see Fig. 1.6). Covering essentially the same area as the radio image in Fig. 5.7, this spectacular X-ray view of the galaxy and its jets has not only confirmed the high-energy activity associated with the powerful expulsion of relativistic plasma, but has also provided scientists with some surprises as well. The X-ray and radio profiles of the jet are quite different; the former is much more uneven than originally anticipated. These results indicate that the energetic particles ejected from the active nucleus do not glow continuously along the whole structure. (Image courtesy of R. Kraft *et al.*, and NASA/Smithsonian Astrophysical Observatory)

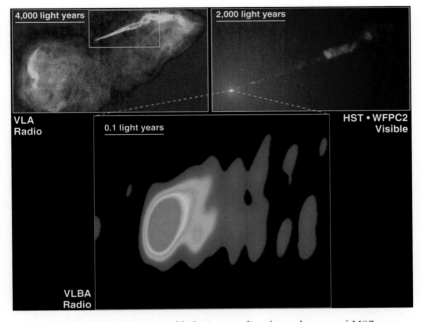

Figure 5.10 The "ray of light" protruding from the core of M87 was noted by astronomers in the early 1900s. Streaming out from the center of this galaxy like a cosmic searchlight is one of nature's most amazing phenomena — a black-hole-powered jet of sub-atomic particles traveling at nearly the speed of light. The source of this grand display is a powerful central object with a mass of 3 billion Suns, 1000 times bigger than the black hole at the center of our Galaxy. The light that we see is produced by electrons twisting along magnetic field lines in the jet, a process that gives it the bluish tint in the upper-right-hand image. This sequence of photographs shows progressively magnified views: *top left*: a VLA image showing the full extent of the jets and the blobs at the termination points; *top right*: a visible light image of the giant elliptical galaxy M87, taken with NASA's Hubble Space Telescope; *bottom*: a Very Long Baseline Array (VLBA) radio image of the region surrounding the black hole. The VLBA is a network of ten radio telescopes utilizing very large baseline interferometry to discern fine detail in the source (see Figs 5.4 and 5.5). This breathtaking view shows how the extragalactic jet is formed into a narrow beam within a few tenths of a light-year of the nucleus (the red region is only a tenth of a light-year across). This distance corresponds to only 100 Schwarzschild radii for a black hole of this mass! (Photographs courtesy of the National Radio Astronomy Observatory and the National Science Foundation [top left and bottom] and J. Biretta at the Space Telescope Science Institute, and NASA [top right])

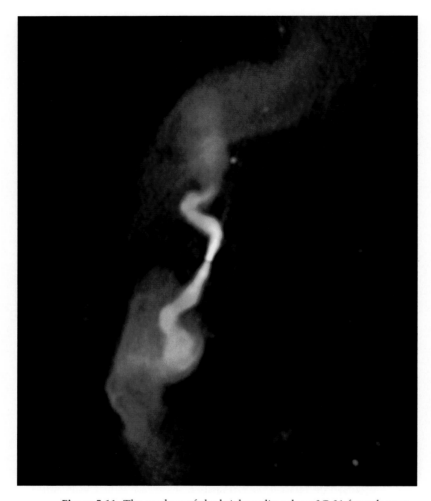

Figure 5.11 The nucleus of the bright radio galaxy 3C 31 funnels powerful streams of energetic particles into the surrounding medium at very nearly the speed of light. The conical jets develop into distorted plumes over an incredible distance of 1 million light-years from the center. But most striking of all is the apparent precession of the "nozzle," which produces the swerving stream of particles evident in projection on this image. Perhaps the black hole is in a binary orbit (like in Fig. 4.5), or the disk of accreting material is periodically realigned. (Image courtesy of R. Laing, A. Bridle, R. Perley, L. Feretti, G. Giovannini, P. Parma, NRAO, AUI, and NSF)

Figure 6.1 NASA's Hubble Space Telescope peered deeply into the most remote recesses of the cosmos and produced this unparalleled view of the visible edge of the universe. Known as the Hubble Deep Field (or HDF), this view has uncovered several hundred never-before-seen galaxies. Besides the classical spiral and elliptical shaped aggregates of stars, it contains a bewildering variety of other galaxy shapes and colors that bear on the structure of the early universe. Some of the galaxies seen here may have formed less than 1 billion years after the Big Bang. This image shows about a quarter of the full HDF view, which was selected for deep exposure because it is ``contaminated'' by only a few foreground stars in our Milky Way along this line of sight – the relatively bright object with diffraction spikes just left of center is one of them. Instead, the vast majority of objects seen here are distant galaxies, some more than 10 billion light-years away. (Image courtesy of R. Williams and the Hubble Deep Field Team at the Space Telescope Science Institute, and NASA)

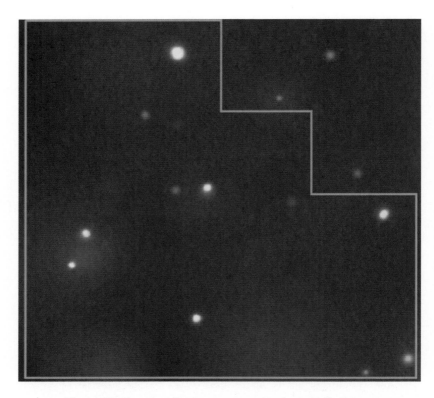

Figure 6.2 Covering the very same portion (essentially the segment to the left of the staggered line) of the sky as that of the Hubble Deep Field, a quarter of which is shown in Fig. 6.1, Chandra's version of the HDF, seen here in false color, probes the glow of X-rays from the farthest reaches of the known universe. Twelve X-ray sources stand out. The hues were selected to represent ``X-ray color,'' in which red denotes ``cooler'' X-rays, while blue corresponds to ``hotter'' X-rays. About half of these sources are due to accretion onto supermassive black holes. The rest are fairly nearby. Besides accretion onto supermassive objects, supernova remnants and neutron stars also contribute to the overall X-ray haze seen in this image. Chandra is now probing far enough into the universe to detect the type of X-ray emission that one finds in galaxies such as the Milky Way, only much farther back in time, to see what our own galaxy and neighborhood might have looked like shortly after the Big Bang. (Image courtesy of G. Garmire, N. Brandt, *et al.*, and NASA/Pennsylvania State University)

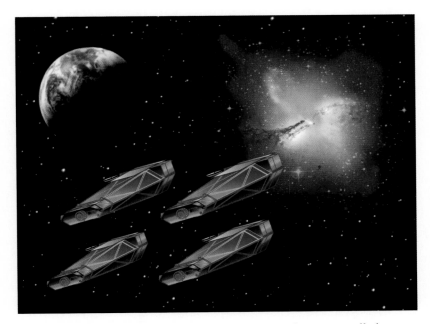

Figure 6.3 The plan for the Constellation-X Observatory calls for a combination of several X-ray satellites orbiting in close proximity to each other. Working in unison to generate the observing power of one giant telescope, they will be able to collect more data in one hour than current X-ray telescopes collect in days or weeks. In combination, the four instruments of Constellation-X will have a sensitivity 100 times greater than any past or present X-ray satellite mission. They will be able to identify thousands of faint X-ray emitting sources, not just the bright ones available to us today. The X-ray source glowing in the distance is a simulated view of Centaurus A (see Figs 1.6, 5.7, and 5.8). (Image courtesy of NASA)

molecular gas, each measuring hundreds of light-years across, will get compressed and a burst of new lights will illuminate previously darkened patches of the cosmos. The sky will grow increasingly jumbled with tattered lanes of dust and gas interspersed among the millions of brilliant young stars and clusters. The beautiful spiral disks that for billions of years will have defined the lanes of our galaxy and those of Andromeda will begin to disintegrate under their mutual gravitational influence.

About 40 million years into the collision, a distant observer in some far-off galaxy will see the Milky Way and Andromeda galaxies very much as we now view NGC 2207 and IC 2163 (see Fig. 4.7). These two bodies are well into their encounter, as evidenced by the latter's tidal tails sweeping stars and gas out 100 000 light-years toward the right-hand side. Computer simulations of this encounter, carried out by a team led by Bruce Elmegreen at the IBM Research Division and Debra Elmegreen at Vassar College, indicate that IC 2163 is swinging past NGC 2207 in a counterclockwise direction. Like the future collision between our galaxy and Andromeda, however, the interaction is not yet over at this stage of the proceedings, for IC 2163 does not have sufficient energy to escape from the gravitational clutches of its partner, and is destined to be pulled back toward the larger galaxy again in the future. Trapped in their mutual dance, these two galaxies will continue to distort and disrupt each other until, billions of years from now, they will merge into a single, more massive structure.

As they swing past each other, the Milky Way and Andromeda will survive their initial collision (see Fig. 4.8), though an inexorable sequence of distortions to their shape will already have begun. After grazing by our galaxy, Andromeda will take perhaps another 100 million years to execute a slow graceful U-turn, before plunging nearly head-on back into the heart of the Milky Way. Newly created sweeping spiral arms will fling stars and gas into the intergalactic medium, preserving only the galactic cores for a future round of convulsions (see Figs. 4.9 and 4.10). With each bounce, an even more spectacular burst of star formation will then occur, while the efflux from supernova explosions will drive much of the remaining gas and dust out of the remnant. Eventually, after 1 to 2 billion years of this drama, the stars from the two galaxies will intermingle to form a single elliptical galaxy (see Fig. 4.11).

Any hint of the Milky Way and Andromeda as majestic spiral galaxies will be gone, and the view our descendants will have should depend on how the Sun fares. There are two possible fates for our solar system, depending on where it is located along its galactic orbital trajectory at the time of the collision. In the first case, the Sun may be caught on a tidal tail and surf its way into the darkness of intergalactic space along with millions of other unfortunate stars. Should this occur, our solar system would eventually find itself very isolated, surrounded by few stellar neighbors, and the night sky would be very dark, sprinkled only with fading embers.

In the second case, the Sun instead finds itself catapulted into the center of the merging traffic, where a great burst of new star formation will be underway. The rate of supernova explosions in this milieu will be very high at first, perhaps as many as a few per year. So the nighttime sky should be aglow with the faint light of fading supernova shells. Conventional ground-based astronomy would be difficult under such circumstances, but one doubts that this archaic pursuit will be necessary by that point! Still, future astronomers would gaze out onto a starry vault and peer all the way into the core of the newly minted elliptical galaxy. There would be no indication that once there were two separate spiral structures threatening to dance the night away, one called the Milky Way and the other Andromeda by a long forgotten civilization.

Who knows? By that time astronomers may not have to look out to the edge of the universe anymore to see quasars in full flight. The black hole at the center of our Galaxy has a mass of about 3 million Suns; its partner in the nucleus of Andromeda is ten times bigger. Their coalescence and subsequent gleeful accretion of the turbulent gas churning in the middle of the merger remnant will likely produce a gargantuan object with the mass of 100 million Suns. And so the story of evolution in the universe will continue. Galaxies merge, supermassive black holes grow in their crucible of dazzle and panic, and this one – the one yet to be created in our midst – will ignite with the power of its brethren now illuminating the universe from the beginning of time. In 5 billion years from now, our descendants may have to look no farther than the center of the new galaxy, perhaps only 50 000 light-years away, to see a quasar resplendent in its youthful exuberance.

4.6 MIDDLEWEIGHT BLACK HOLES

We will end this chapter on a somewhat speculative note, bearing on the very latest investigations now being carried out by serious black-hole hunters. Though the shift in paradigm concerning the existence of supermassive objects has been overwhelming and virtually complete, astronomers have always felt uncomfortable with the idea that black holes seemingly come in only two distinct and unrelated categories – those produced in supernova explosions, weighing a mere 10 to 20 solar masses, and the million- to billion-solar-mass behemoths we have been considering in this book. General relativity places no limit on the mass of these objects, so where are the intermediate-size black holes?

Well, scientists are now realizing that middleweights may have already been found years ago, though lacking proper identification; their existence is still unconfirmed and hotly debated. When first discovered, the so-called ultraluminous X-ray sources elicited only mild interest and interpretative analysis. In the context of stellar-size objects, however, their brightness is peculiar because 10-solar-mass black holes pumping out too much radiation prevent the surrounding matter from accreting onto them – a cosmic version of "biting the hand that feeds you." But this problem can easily be alleviated if the underlying objects are bigger, since strong gravity can overwhelm the expulsive power of radiation. If correct, this interpretation suggests that ultraluminous X-ray sources are intermediate-size black holes with a mass of 100 to 1000 Suns. Countering this argument, astronomers such as Andreas Zezas of the Harvard-Smithsonian Center for Astrophysics, suggest that these may be small objects after all, simply pumping all their X-rays out into narrow cones, making them look brighter than they really are.

Understandably, this debate is more than an idle exercise, since middleweights could in principle merge and grow to become supermassive black holes in their own right. In a recent survey conducted with the German-led ROSAT X-ray satellite program, Andrew Ptak of Carnegie Mellon University and Edward Colbert of Johns Hopkins looked for these sources in 750 galaxies within 200 million light-years of Earth and found that roughly one in 35 of them possesses objects as massive as 100 Suns, some higher.[20] But they are typically found away from the core of the parent galaxy, whereas heavyweights always lurk right in the middle.

NASA's Chandra X-ray Observatory has taken up the challenge of better defining these peculiar objects, and in the past several years has uncovered another interesting property. It seems that "starburst" galaxies have a proportionally higher number of intermediate-size black holes than do other galaxies. As their name suggests, starburst galaxies are characterized by an unusually high rate of star formation, endowing them with an uncommon brightness caused by a high concentration of young, massive stars and supernova explosions. Under such ideal conditions, astronomers believe, numerous 10- to 20-solar-mass-size black holes should emerge – and ultimately merge – to account for the higher incidence of middleweights in these systems.

More exciting still was the discovery that in at least one such galaxy (see Figs. 4.12 and 4.13), several mid-size black holes appear to be *sinking* toward the nucleus. The galaxy NGC 253, a relatively nearby classical starburst galaxy about 10 million light-years from Earth, gave no prior indication of black-hole activity in its central region. When Chandra focused on its nucleus, however, it uncovered

---

[20] The comprehensive list of intermediate-size black holes surveyed by ROSAT appears in Colbert and Ptak (2002).

10 ultraluminous point sources, three of which are located within 3000 light-years of the middle (see Fig. 4.13). To have three of these objects confined to such a relatively small volume is very unusual, suggesting that they may have migrated there from creation sites farther out.[21]

Thrashing to the center of NGC 253, these three middleweights may someday coalesce with the central black hole inveigling them to join in, and then continue to grow even more by accreting matter from the surrounding medium. In other words, this starburst galaxy may be transforming itself into a quasar-like object right before our eyes. But we will have to wait for farther developments to be certain that ultraluminous X-ray sources are indeed intermediate-size black holes.

[21] This work was carried out by Kimberly Weaver of NASA's Goddard Space Flight Center, and David Strickland and Timothy Heckman of Johns Hopkins University. The team observed NGC 253 for 3.5 hours using Chandra's Advanced CCD Imaging Spectrometer.

5   Relativistic ejection of plasma

Other than the spectacle of an obscured event horizon quivering before a bright sheet of background light, the most spectacular black-hole phenomenon astronomers can witness from the remoteness of Earth is a relativistic jet of plasma piercing the darkness of inter-galactic space. Among the most dizzying cosmic displays in nature, these funnels of energetic particles probe the medium surrounding roughly one in 20 known supermassive black holes. A prominent jet was evident on the very first quasar photograph (of 3C 273), and glows even more brilliantly as a high-energy ray of light in modern Chandra images (see Fig. 1.2). For the most part, however, black-hole jets manifest themselves in a "parallel" universe – indeed, their ghostly apparitions pre-empted the discovery of supermassive black holes by several decades, though without any portent of what they would later reveal. And once again, astronomers can thank the telephone company for facilitating one of the most amazing advances in the history of science, on a par with the discovery – six decades later – of the cosmic microwave background radiation through the commercialization of space.

Not long after a demonstration that the substance of light behaves like a series of waves undulating through time and space, Guglielmo Marconi (1874–1937) successfully initiated transatlantic communications in 1901 using wireless radio. The commercial use of this new technology flourished; radio equipment was installed on ships – including the RMS Titanic, whose tragedy would have been greater had not wireless communication with nearby vessels effected the rescue of 705 passengers – and large telegraph-like companies were created to dispatch a burgeoning torrent of information from continent to continent. The telephone company realized that shortwave links, operating in the wavelength range of 10 to 100 meters, could be used to carry intercontinental phone calls without the associated expense of laying cable across the ocean floor. But as any ham or shortwave listener knows only too well, shortwave communication is annoyingly hampered by too much static and noise.

At AT&T's Bell Labs in New Jersey, the young radio engineer Karl Guthe Jansky (1905–50) was assigned the task of identifying the sources of shortwave noise that might interfere with radio voice transmissions (see Fig. 5.1). So in 1931 he built a highly directional antenna to receive signals with a wavelength of about 14.5 meters, and mounted it on a turntable (known whimsically as "Jansky's merry-go-round") that could point in any direction (see Fig. 5.2). After months of systematic observations, he identified three types of static: nearby thunderstorms, distant thunderstorms, and a faint steady hiss of unknown origin.

He spent over a year trying to figure out the nature of the third type, which rose and fell once a day. Jansky thought for a while that he was seeing radiation from the Sun but, after a few months, he could see that the brightest point in his signal kept moving. There was also that curious and nagging cycle of repetition – the signal rose not every 24 hours, but every 23 hours and 56 minutes. As most amateur astronomers know, this is a clue that the source of the radiation is fixed to the stars rather than the Sun, because Earth's advance along its orbit causes points in the firmament to cross the meridian earlier by 4 minutes every day.

As it turns out, Jansky had discovered radiation concentrated in the constellation of Sagittarius, along the Milky Way. The origin of this peculiar radio noise apparently had something to do with the galactic center, which was already known by then to reside in this portion of the sky, and his surprising announcement was reported on the front page of the *New York Times* on May 5, 1933. Yearning for a deeper understanding of this mysterious radio glow, Jansky implored Bell Labs to build a 30-meter dish antenna, but his company had the

FIGURE 5.1 Karl Jansky, a radio engineer at Bell Telephone Laboratories in Holmdel, New Jersey, built the first "radio telescope" (see Fig. 5.2) to search for sources of interference that would impact the burgeoning field of wireless communications. In the process, he discovered radio waves arriving from outside Earth's atmosphere, with the strongest concentration at the center of the Milky Way galaxy, in the constellation of Sagittarius. (Image courtesy of NRAO, AUI, and NSF)

FIGURE 5.2 Assigned the task of evaluating short waves for use in transatlantic radio telephone communications, Karl Jansky built this antenna to receive radio waves with a wavelength of about 14.5 meters. By mounting it on a turntable, he was able to determine the direction to any signal. In this way, he discovered that the center of our galaxy, in the constellation of Sagittarius, produced the strongest concentration of radio waves, and thereby created the field of radio astronomy. (Image courtesy of NRAO, AUI, and NSF)

answer they needed about the background static, and their young engineer was assigned to another project. The world's first radio astronomer thus ended his scientific career after his very first discovery – but what an achievement that was!

Though most scientists did not at first appreciate the significance of Jansky's demonstration, news finally reached Wheaton, Illinois, where it inspired Grote Reber, another radio engineer, to study cosmic radio waves. In the 1930s, Reber applied for jobs with Jansky at Bell Labs and with assorted astronomical observatories around the world, but the Great Depression was in full swing, and none of them were hiring at the time. So the world's second radio astronomer built his own 32-foot-diameter parabolic dish antenna in his backyard, while working full time for a radio company in Chicago.

Reber found that sparks in automobile engines created too much interference during the daytime; he therefore spent long hours scanning the heavens at night, and finally produced the first survey of the sky at radio wavelengths in 1941. The great advances in microwave technology that had produced radar during the war became available to astronomers soon after that, and the exploration of the radio universe – that hidden reality existing in parallel to the one that connects to our senses – blossomed thereafter.

Showcasing his data as contour maps delineating regions of varying intensity, Reber not only confirmed Jansky's discovery that the brightest part of the radio sky lies toward the center of the Milky Way in the southern hemisphere, but he also uncovered equally astounding bright radio sources in other constellations, including one in Cygnus. We know it today – with immeasurably higher resolution – as the finest manifestation of relativistic plasma ejection by a supermassive black hole (see Fig. 1.8).

This discovery of a "parallel" universe was the élan vital for the young field of radio astronomy, whose rapid growth could be measured by the increasing number of catalogs bursting with ever-fainter radio sources that could be observed with improving instruments and sensitivity (see Fig. 5.3). Not all the objects we see in the night sky emit radio light, nor have all the radio sources detected over the years been confirmed on optical images. But though the overlap between the two cosmic maps is incomplete, cross-registration of objects in several categories is now possible. One of the more significant successes in this collation also happens to have been one of the very first – Cygnus A was identified with what were initially thought to be two colliding galaxies in 1952, providing not only evidence that this source was extragalactic, but also that some nonstellar mechanism must be responsible for the prodigious outpouring of radio waves emanating from the middle of this fracas.[1] Since then, numerous similar jet sources have been uncovered and, in this chapter, we shall focus on the peculiar physical principles they embody.

5.1   IMAGING SUPERMASSIVE BLACK HOLES

The luminous extensions in Fig. 1.8 project out from the nucleus of Cygnus A an incredible distance three times the size of the Milky Way. Yet located 600 million light-years from Earth, they cast an aspect only one-tenth the diameter of the full moon. The challenge for radio astronomers, therefore, is to create the eye-pleasing resolution evident in this photograph, given the additional complication that radio wavelengths are almost a million times longer than those of visible light. A telescope's ability to see fine detail in the target depends primarily on how many wavelengths it can squeeze into its aperture. So a radio dish with even modest capability must be extremely large to achieve the same angular resolution as a small optical telescope – essentially *thousands* of kilometers. Mechanically, however, such a large structure cannot be engineered. How, then, do radio astronomers get around this problem and expose the invisible radio universe? A clever solution was found with the development of radio "interferometry."

Rather than imaging a celestial object with a single giant telescope, the technique of Very Long Baseline Interferometry (VLBI) utilizes an array of smaller radio dishes dispersed over a very large area. The radiation from the source is detected at slightly different times in each receiver, according to its position on the Earth or in space. But the signals can be combined with a powerful central computer, permitting the network to function as just one instrument with an equivalent size the distance between the various components participating

---

[1] After the Cambridge astronomer F. Graham Smith managed to position Cygnus A accurately, Walter Baade identified it with a faint optical source that he interpreted as being two colliding galaxies. A more recent interpretation holds that it is a giant elliptical galaxy whose body is bifurcated by a dust lane left over from a spiral galaxy that it recently swallowed. See Osterbrock (2001).

in the observation. The collected information is recorded on magnetic tape and shipped to the control center for processing at a later date. It is therefore critical for this system to be driven by the most accurate timing device possible, since the different signals must be synthesized with precise knowledge of the radiation's arrival time at each location. That function is now assigned to a highly stable hydrogen maser clock.

In the late 1970s, the Very Large Array (VLA), the workhorse of high precision radio astronomy over the past several decades, was built near Socorro, New Mexico. Its Y-shape consists of 27 individual antennas (see Fig. 5.4), and the data from all its components can be combined to accomplish the resolving power of a single dish 36 kilometers in diameter!

Increasing the baseline (i.e., the effective "aperture" of the network) was the main motivation behind the construction 15 years later of the next system – the Very Long Baseline Array (VLBA) – by the National Radio Astronomy Observatory and Massachusetts Institute of Technology's Haystack Observatory. This facility consists of ten radio telescopes spread over 8000 kilometers across the surface of the Earth, from Mauna Kea in Hawaii to St Croix in the Virgin Islands (see Fig. 5.5). With the power to simulate a single antenna the size of our planet, this technological wonder has the power to see features as small as the Earth's orbit at the galactic center, some 28 000 light-years away, and an object as small as our solar system in Andromeda, a hundred times more distant.

The astounding capability of the VLBA was illustrated by its observation of the galaxy NGC 4258 (see Fig. 2.6), some 23 million light-years away, which produced the most elegant and compelling evidence so far for the existence of extragalactic supermassive black holes. Using this instrument to measure the velocity of water molecules at the nucleus of NGC 4258, astronomers have determined that 40 million Suns' worth of material is concentrated within a radius of less than half a light-year, currently the highest *confirmed* large-scale density of matter in the universe.

Today, the joint efforts and signals of the VLA and VLBA can be combined with those of the orbiting Japanese satellite HALCA (the Highly Advanced Laboratory for Advanced Communications and Astronomy) to triple the resolving power previously available with only ground-based telescopes. The resulting Space VLBI satellite system is 100 times more powerful than the Hubble Space Telescope. In fact, its resolving power is equivalent to being able to see a grain of rice in Tokyo from a perch in Los Angeles.

But many objects in space require still higher resolution than this and the field of radio astronomy has been mobilized to provide greater capability in the near future. Researchers at the Jet Propulsion Laboratory in California are planning a project known as the Advanced Radio Interferometry between Space and Earth (ARISE), with a possible launch date of 2008. The mission calls for the launch of two 25-meter telescopes in very elliptical orbits – to maximize the coverage over a broad range of baselines – simulating a telescope four times bigger than our planet. Its resolving power will permit astronomers to see objects only ten times bigger than the Earth's orbit in Andromeda. But most spectacular of all will be its capability to "see" the structure of an accretion disk wrapped around the supermassive black hole at the galactic center. Indeed, with this type of resolution, ARISE may be able to image the black hole itself, whose event horizon is three times bigger than the smallest region viewable with this facility of the future.[2]

5.2 JETS FROM SUPERMASSIVE BLACK HOLES

The very idea of supermassive black holes launching plasma into intergalactic space seems antithetical to our perception that their power of attraction is overwhelming and inexorable. A single particle, after all, would always lose its duel with the ponderous object and succumb to its entombment below the latter's event horizon. Yet when matter falls toward the black hole in concentration, astronomers often

---

[2] For a comprehensive discussion of this exciting, future development, see Melia (2003).

see an ensuing fusillade of scorching gas cavitating long narrow funnels in the medium surrounding the host galaxy. Furthermore, these energetic streams appear to be a dynamically significant influence – they're not merely harmless apparitions pointing their accusatory fingers back to where the launch took place. In Cygnus A, for example, the radio specter of this commanding source (see Fig. 1.8) is eclipsed by an even more dramatic image in X-rays, showing the jets' deleterious impact on an otherwise quiescent primordial substrate (see Fig. 5.6). The phenomenon of black-hole gas expulsion generates considerable excitement among the astrophysics community because it produces the fastest moving plasma in the universe. The challenge is to understand why this relativistic acceleration even occurs.

When, in addition, scientists view the enormous cavity carved out of the universe by the energetic expulsions in Cygnus A, they are compelled to acknowledge the sobering fact that this structure has been maintained for at least as long as it takes the streaming particles to journey from the center of the galaxy to the extremities of the giant lobes. In other words, these pencil-thin jets of relativistic plasma have retained their pristine configuration for over one million years! It is not surprising, therefore, that the most conservative view now maintained by physicists is that a spinning black hole is lurking at the nucleus. The axis of its spin functions like a steady rudder, an immovable gyroscope, whose direction pre-determines the orientation of the jets. No one has produced an alternative physical description of how such a large, steady structure could otherwise be maintained. Although the definitive mechanism for how the ejection takes place is still to be worked out in detail, we shall see below that mounting evidence now points to the twisting motion of magnetized plasma near the black hole's event horizon as the cause of the expulsion. The Kerr spacetime, which describes the dragging of inertial frames about the black hole's spin axis, provides a natural setting for establishing the preferred direction for this ejective process.

Much closer to home, Centaurus A (11 million light-years distant versus 600 million light-years for Cygnus A) is not quite as powerful as Cygnus A, but the two are similar in other ways, for example in producing twin jets of material moving close to the speed of light over unimaginably great distances. Centaurus A (see Fig. 1.6) is too far away for us to discern yet what fraction of dark matter in its nucleus is due to dead stars as opposed to a single object, but the indications are that a supermassive black hole, with a mass possibly 100 times larger than that of Sagittarius A* at the galactic center, is responsible for all the nonstellar activity (see Fig. 1.7). Aside from the compelling arguments we considered in Chapter 1, the existence of jets emanating from the nucleus of this galaxy – like those produced by Cygnus A – would otherwise be very difficult to explain (see Fig. 5.7).

The radio power of a galaxy such as Cygnus A is 10 to 100 times greater than the combined light of the hundreds of billions of stars within its girth! When giant lobes are evident, they account for most of this output, though often the trails of high-velocity particles and their associated fields that energize these cosmic "puffs" of glowing gas are also visible on radio maps. It was recognized early on[3] that the radio waves seen at the Earth are undoubtedly synchrotron radiation, produced when highly energetic electrons (those traveling at nearly the speed of light) emit light as they gyrate wildly in magnetic fields. The radiation is essentially "torn" away from these particles as they accelerate back and forth, much as you would lose your hat and scarf and other loosely attached items of clothing if you were whipped around on a merry-go-round spinning faster and faster.

But from the very beginning there was something very peculiar about the core in certain radio images. It turns out that this central spot, coincident with the supermassive black hole, is sometimes too small to possibly radiate all the radio waves detected from it by astronomers. Although the radiation could in principle be produced within such a tiny region, it would then not be able to escape through the highly condensed particles. The resolution to this problem

---

[3] Some of the pioneering efforts in this regard were due to Fred Hoyle, Geoffrey R. Burbidge, and their collaborators, whose papers appeared in *Nature* and the *Astrophysical Journal* toward the end of the 1960s and in the early 1970s.

eventually produced one of the earliest breakthroughs in our understanding of jets and their behavior. Suppose the hot, magnetized
plasma was actually moving at a very high speed toward Earth,
suggested Roger Blandford and Martin Rees in 1978. Then beaming
effects, due in part to strong Doppler boosting, would produce a significantly higher radiation intensity for any observer happening to lie
within the cone of the jet than for one located outside.

This interpretation has become the paradigm for how tiny cores
can appear to be so bright across such vast expanses of the universe.
Quasars and the brightest nuclei of galaxies hosting supermassive
black holes are precisely those jet sources whose searchlight beams
of energy and light are directed right toward us. The success of this
explanation in resolving the mystery of overly bright, tiny cores also
constitutes the first piece of evidence that the plasma within the jet is
moving very rapidly – typically at 99.99 percent of the speed of light!

Once the Very Large Array began its systematic study of radio
jets, the floodgates of new and comparative information opened wide.
This telescope array has the sensitivity to detect even weak jets very
quickly. It also has the dynamic range to do so in the presence of bright
unresolved background light, and is big enough to separate out the
many components often seen inside the streams of plasma. Examples
of jets in all types of radio-emitting active galactic nuclei began to
surface, and the identification of the pathways along which energy
was transported to the outer, bright lobes gave credence to the picture
painted by physicists trying to explain the jet phenomenon.[4]

Today, astrophysicists studying jets concern themselves with
two principal issues – how is the plasma expelled by the black hole
(addressed in the next section) and how do they acquire their internal structure? Radio images of the brightest, well-resolved funnels

---

[4] This theoretical interpretation had its roots in early papers by Morrison (1969), who
attempted to explain jets as relativistic beams of particles and energy, and by Rees
(1971), who suggested that the radio-emitting expulsions were powered by low-
frequency electromagnetic beams. Somewhat later, Longair *et al.* (1973) proposed
an energy transport time scale "comparable with the age of the source," and Scheuer
(1974) undertook a dynamical study of radio sources powered by relativistic beams.

reveal a rich variety of internal knots, rings, loops, filaments, and some helix-like "threads." X-ray images of these extensions only complicate matters even more. For example, a recent Chandra observation of the jet in Centaurus A (see Fig. 5.8) has produced quite a few surprises. The X-ray and radio structures of the plasma funnels are significantly different, and the X-ray jet is much more uneven than originally believed. These results have cast some doubt on the simplest picture of how the energetic particles expelled from the active nucleus travel along the jet. Almost certainly, some measure of instability must be the cause, since nature tends not to maintain "perfect" conditions for very long.

At any rate, our chief concern in this book is what these jets have to say about the nature of supermassive black holes. Astronomers now realize that the plasma is expelled at nearly the speed of light. Not too long ago, they acquired some evidence that the composition of this screeching gas may be a mixture of matter and antimatter – basically electrons and positrons, each the other's antiparticle – rather than ordinary matter composed of dissociated hydrogen.[5] Radio waves behave differently when they propagate through the former compared with their passage through the latter. In at least one case – the quasar known as 3C 279 – this difference is measurable, and matter–antimatter has won the verdict. Scientists also know that jets are sustained over millions of years, so explosive events are ruled out as the cause of their expulsion. And finally, in at least 30 known cases, some features in the jets are moving across the sky *faster* than the speed of light. Let us see now how all of these clues appear to have ultimately melded into a coherent explanation.

5.3 FASTER THAN LIGHT MOTION

In the so-called superluminal sources,[6] the streams of particles move at relativistic speeds away from the center of some galaxies and

---

[5] This inference was based on the observed polarization properties of the quasar known as 3C 279 by Wardle *et al.* (1998) at Brandeis University.

[6] See Zensus and Pearson (1987) for a comprehensive compilation.

produce features that drift across the heavens with velocities significantly greater than that of light – a phenomenon that caused an understandable stir when it first became known. The public reaction to the discovery of this apparent motion was generally one of skepticism, inducing some to refute special relativity and/or the concept of cosmological redshifts.[7] To understand what is happening, and to uncover what the latest observations are now telling us about the acceleration of jets, let us consider one of the most spectacular members of this class of objects – the enormous elliptical galaxy known as M87 (see Fig. 5.9).

The Milky Way, together with Andromeda and their entourage of smaller bystanders, orbit on the outskirts of the Virgo cluster, a truly gigantic assemblage of thousands of galaxies, of which M87 is the largest. Discovered by the French astronomer Charles Messier in 1781, M87's optical jet was first seen in 1918 by Heber Curtis[8] (1872–1942) of the Lick Observatory, who described it as "a curious straight ray." M87 is in fact the first extragalactic object for which the term "jet" was invoked. In 1954, Baade and Minkowski described this optical feature as "a unique peculiarity known for a long time . . . a straight jet extending from the nucleus in p.a. 290°, bluer than the nebula itself . . . several strong condensations." Nine years later, Schmidt's (1963) identification of "a star of about thirteenth magnitude and a faint wisp or jet" near the accurate positions of the radio components of 3C 273 provided the first indication that "jets" also produce radio waves.

---

[7] For an example of the discussion during this debate, see Stubbs (1971).

[8] As an interesting historical aside, Curtis was one participant in the now-famous Shapley–Curtis debate in which the featured topic was the question of whether the universe consisted of just a single giant galaxy (Shapley's position) or whether the Sun was situated near the middle of the Milky Way, a relatively small galaxy among many. As we saw earlier in this book, a partial resolution to the debate came in the 1920s, when astronomer Edwin Hubble successfully showed that Andromeda was much further away than the size even Shapley had attributed to the universe. Of course, the galaxy turned out to be much bigger than Curtis had allowed, and the Sun orbits well away from its center, but he was correct in his early assessment of the multi-galactic structure of the universe.

FIGURE 5.9 The giant elliptical galaxy M87 is the dominant "gorilla" at the center of the Virgo cluster, to which the Milky Way is loosely attached. M87 has a diameter of half a million light-years which, at a distance of 60 million light-years from Earth, subtends an angle of over half a degree – more than the diameter of the full moon. Its outer layers appear noticeably distorted, probably because of their gravitational interaction with other Virgo cluster members, and because they contain remnant material from disrupted galaxies that have merged with M87 during close encounters in the past. To the bottom right of the galaxy, we can just barely make out a jet of material discovered in 1918 by H. D. Curtis of the Lick Observatory. Seen much more prominently at other wavelengths (see Fig. 5.10), this pencil-thin beam of relativistic particles displays apparent superluminal motion, which is best understood in terms of rapid advancement in a direction close to the line of sight. (Image courtesy of NOAO/AURA/NSF)

On very deep photographic plates, M87 extends out over half a degree – more than the diameter of the full moon and, at a distance of 60 million light-years, this corresponds to a linear extension of about half a million light-years. It appears noticeably distorted on the fringes, indicating a significant gravitational interaction with other galaxies in the Virgo cluster, and possibly the result of recent mergers with smaller aggregates of stars. Filling a much larger volume of space than the Milky Way, it contains many more stars (and a much greater mass) than our galaxy, certainly numbering in the several trillion. Following convention, in which Centaurus A and Cygnus A were named as such because they are the brightest radio sources in their respective constellations, M87 is also known to radio astronomers as Virgo A.

M87 has become one of the most readily chosen objects for study because it is one of the nearest jet-producing galaxies and its strong radio emission makes it an excellent target for radio telescopes. It is also very alluring because of what lurks in its interior. Repeating the work carried out for Centaurus A – in which the Hubble Space Telescope identified in spectacular fashion not only the point source associated with the supermassive black hole in its nucleus (see Fig. 1.7), but also its mass – for the nucleus of M87, astronomers have now shown that a dark mass of 3 billion Suns (about 1000 times bigger than Sagittarius A* at the galactic center) is concentrated into a volume no larger than our solar system. And focusing in on this exotic site, astronomers can now conduct the most up-to-date high-resolution radio measurements, revealing what is happening a mere one tenth of a light-year from the nucleus – a size no bigger than 50 times the diameter of the supermassive black hole (see Fig. 5.10). They are literally witnessing how these splendid jets form within sight of the event horizon itself.

One of the more breathtaking results of this groundbreaking work is evident in the lower panel of Fig. 5.10, produced with the Very Large Baseline Array by a group of astronomers at the Space Telescope

Science Institute and the University of New Mexico.[9] The sheer size of this supermassive black hole, and its relative proximity to Earth compared to quasars and many other active galaxies, is giving scientists an unparalleled view of the mysterious region where the powerful stream of subatomic particles is accelerated to near lightspeed. These astronomers had speculated that the jet ought to be launched near the black hole and that one should see some evidence of the active mechanism at work, but as they looked closer and closer to the center, reaching a distance of less than 50 times the diameter of the event horizon, they kept seeing an already-formed beam.

According to the highly detailed VLBA image shown in Fig. 5.10, M87's jet adopts its narrow elongated shape within a mere fraction of a light-year from its point of launch. At the very base, sampling regions within only a few *hundredths* of a light-year of the black hole, the investigators were able to ascertain that the inner portion of the outflow is instead very broad. Its opening angle of about 60 degrees is much wider than anyone had anticipated. In other words, contrary to everyone's expectation that the jet should begin its outward excursion as a relatively narrow funnel centered on the black hole, it instead appears to be bubbling upwards from a much wider region – undoubtedly from the accretion disk wrapped around the center. Clearly, the black hole is not acting alone.

Astrophysicists now believe that many aspects of quasar jets may be understood in the context of relativistic flows launched by what they call the "B++" mechanism – a *b*lack hole *plus* a rotating accretion disk *plus* a magnetic field that is anchored in this disk and wound up by its rotation. Think of the magnetic field as wires threading the plasma, both descending deeper and deeper into the gravitational well of the black hole. The infalling gas is not only compressed – which at the same time increases the density of wires – but it also

[9] The investigators in this project were John Biretta and Mario Livio of the Space Telescope Science Institute in Baltimore, Maryland, and William Junor of the University of New Mexico, in Albuquerque. Their findings were published in 1999 by *Nature*.

rotates around the central object faster and faster as it gets closer. The wires get braided and twisted, and some break. The electrically charged particles can flow like beads *along* the wires, but not across them. What we are seeing in the bottom panel of Fig. 5.10 is this profuse efflux of charged particles abandoning the disk as the twirling wires fling them outwards and upwards. This partially explains why the jet appears to be "already formed" all the way down to the smallest distances sampled by the VLBA observations, because the plasma is presumably expelled from throughout the disk, not just the black hole itself.

In retrospect, a disk origin for quasar jets is perhaps the main reason why one often sees evidence of precession over their million-light-year excursion into the relative void of intergalactic space. Unless the black hole is in a binary (like Fig. 4.5), it ought to be very stable. Its disk, however, much lighter and subject to the vagaries of plasma ploughing into it from outside, could very well be wobbling over time. But since the braided field lines are more or less perpendicular to its surface, the direction into which the jet is launched may itself precess, producing a corkscrew pattern lasting eons (see Fig. 5.11).

The possibility of probing so deeply into the inner workings of M87's core is one of the reasons why astronomers consider this investigation to have been a spectacular success. But perhaps the best reason to marvel at this unusual galaxy is that the motion of material near the base of its jets has been measured independently with radio telescopes (principally the VLBA) and the Hubble Space Telescope, and they all reveal that the ejected plasma is receding from the supermassive black hole at six times the speed of light. It was in fact this type of situation with quasars back in the 1960s and 1970s that motivated the need for much higher resolution in the observations to see what was happening near the central object. The desire to overcome the observational shortcomings at that time brought together groups from Canada, the USA, the UK, and the Soviet Union, with the goal of developing Very Long Baseline Interferometry (VLBI), in which a global network of independent radio telescopes could merge their detected signals to provide exquisitely finer detail in the source than any of them could produce individually.

The phenomenon of superluminal motion is intriguing, but it does not really have to do with the actual propagation of matter at these velocities. It is an artifact of the finite *and* constant speed of light. Back in 1966, Martin Rees proposed that these effects might be associated with relativistically expanding shells of matter. In later refinements of this basic picture, it became clear that the underlying sources for the observed radiation were quasar jets consisting of beams of relativistically moving particles.[10] The superluminal effect, it turns out, can be understood comfortably *within* the confines of special relativity.

It has to do with the fact that when a source of light is moving toward us it catches up with the radiation it is emitting, so that changes we see in its complexion appear to be happening over a smaller interval of time. Since it then appears that the distance was covered over a shortened duration, we infer a greater velocity than the object actually possessed. Astronomers now understand that the jet in Fig. 5.10, and all those like it that exhibit superluminal motion, must be pointed directly at us. When we look at the VLBA image of M87, we are evidently seeing the much longer jet *projected* onto the plane of the sky – an interestingly complementary perspective on the "tiny but overly bright core" problem we broached earlier, whose resolution has us watching the glowing gas perched squarely *inside* its jet.

Consider the following analogy. Suppose Sam and Eva are playing catch across a tall hedge. Sam can see the ball thrown over it, but he cannot see Eva. Eva tells Sam that she will throw two balls, in quick succession, and that based on the interval of time between them arriving in his hands, he should calculate how fast she was moving when she threw them. Sam knows that she is going to run 10 meters between the two throws, but what he doesn't realize is that she has chosen to run straight toward him. Eva begins her run, throws the

---

[10] See Blandford, McKee, and Rees (1977).

first ball, and waits until she has covered 10 meters before throwing the second. Since she has moved in the same direction as the first ball, and partially caught up with it, the time between the arrival of the second ball in Sam's hands and that of the first is shorter than the time she actually waited between the two throws.

The velocity Sam infers for her is therefore greater than what she can muster because he thinks she covered the 10 meters in less time than it actually took her to do so. Of course, this only works because Eva moved toward him. If she had instead been running parallel to the hedge, Sam's measured interval of time would have been the same as hers and they would then have agreed on her velocity. By extension, astronomers infer that the jets of quasars and active galactic nuclei displaying superluminal motion must be directed straight toward us. Alien eyes viewing these beams of plasma from a different angle, however, do not see features in these jets moving superluminally and, from their perspective, the jets extend much farther out than is evident to us back here on Earth.

Alluring and extraordinarily large, jets have understandably served as a principal method of discovery and identification for supermassive black holes since the early 1970s. No one has successfully explained how relativistic beams of plasma could be created and maintained over millions – perhaps billions – of years without the underlying influence of a ponderous, stable body at the nucleus of the host galaxy. As such, the appearance of a jet is at least prima facie evidence that a behemoth lurks nearby. More direct and compelling means of identification, some of which were described in Chapter 2, are now surfacing, and we may even be on the verge of actually "seeing" an event horizon at the center of the Milky Way. Science marches forward on all fronts, and our exploration of the universe at wavelengths other than radio is beginning to produce exhilarating discoveries of its own. We shall examine some of these findings in the next chapter.

6 Supermassive black holes in the universe

Our view of the night sky is a panoply of stars choreographed to the galaxy's spiral melody. A deep exploration of the universe beyond our immediate neighborhood would therefore not be possible were it not for the occasional chance alignment of interstices among these swarming points of light. For ten consecutive days in December of 1995, the Hubble Space Telescope peered through just such a clearing, and produced our deepest ever view of the universe, graced with thousands of galaxies bursting into life at the dawn of time.

6.1 THE HUBBLE DEEP FIELD

Called the "Hubble Deep Field" (see Fig. 6.1), this image contains not only classical spiral and elliptical galaxies, but also boasts a rich variety of other galaxy shapes and colors that hint at the influences governing the evolution of the early universe. Some of these objects may have condensed within 1 billion years of the Big Bang.

Covering a speck of sky only one-thirtieth the diameter of the full moon, the view of the Hubble Deep Field (one quarter of which is shown here) is so narrow that just a few foreground stars in our galaxy are visible. Most of the objects contained within it are instead so distant that our eyes would have to be four billion times more sensitive in order for us to see them without the aid of a telescope. But even though this field is only a small sample of the entire sky, astronomers consider it to be representative of typical galaxy distributions, under the assumption that the universe looks the same in all directions.

What is not yet apparent from this beautiful and haunting image is that, secreted among the many luminiferous islands of stars, are millions of young, vibrant, supermassive black holes exploding into our awareness in other portions of the spectrum. The opening of this keyhole across the heavens provided an unprecedented opportunity for ground-based radio astronomers to point their array of radio dishes toward a portion of the sky unhindered by the contamination of nearby objects. Unlike Hubble's optical view, radio telescopes can peer beyond the obscuring dust clouds and into the hearts of galaxies in the field. By combining the signals received with nine European telescopes forming part of what is now called the European VLBI Network (EVN), scientists at the Joint Institute for Very Long Based Interferometry in Dwingeloo, the Netherlands, and their colleagues around the world, produced radio images of the Hubble Deep Field that are three times sharper than those even from the Hubble Space Telescope itself.[1]

The high-resolution images show that many of the galaxies in the Hubble Deep Field harbor central massive objects. The researchers were surprised by the diversity of hosts for the radio sources, which included elliptical, spiral, and very distant, dust-obscured starburst galaxies. Even more exciting was the realization that many of the radio sources are quite small – less than 600 light-years wide. This clearly shows that the radio emission is generated by processes associated with supermassive black holes, and therefore supports the theory that black holes are linked with the formation of structure in the early universe.

6.2 THE CHANDRA DEEP FIELD

Even these significant results, however, have since been overshadowed by dramatic discoveries made more recently by combining the power of the Hubble and Chandra observatories and several

---

[1] Michael Garrett of the Joint Institute for Very Long Based Interferometry in Dwingeloo, the Netherlands, and his colleagues combined the radio signals from the 100-meter telescope in Effelsberg, Germany, the 76-meter Lovell Telescope in the UK, the 70-meter NASA/Deep Space Network antenna near Madrid in Spain, and six other large radio telescopes located across Europe. Data for these images were archived on high-speed magnetic tape recorders, and later processed by a supercomputer operated by the National Radio Astronomy Observatory in Socorro, New Mexico. This combined system is equivalent to a supersensitive, giant telescope of continental dimensions. The technical aspects of this observation are discussed in Garrett *et al.* (2001).

ground-based optical and infrared facilities. The deepest multiwavelength look ever made of the distant cosmos has shown that black holes of all sizes ruled the early universe, and that they behaved in more varied ways than researchers had expected. Chandra's X-ray version of the Hubble Deep Field is shown in Fig. 6.2. The image in Fig. 6.1 corresponds to the lower-left-hand quarter of the full field sampled by both of these satellite observatories.

One of the groups conducting this investigation, led by Riccardo Giacconi, co-winner of the 2002 Nobel Prize in Physics, reported that some 350 supermassive black holes appeared in the patch of the cosmos they surveyed. Extrapolated to the whole sky, this would amount to 200 million supermassive black holes spread throughout the early universe! These objects were evidently much more active in the past than they are in the present.

Other groups of astronomers have taken a different, complementary approach to this search, by focusing not so much on all the distant X-ray sources they can find in one patch, but rather by studying the high-energy characteristics of active galactic nuclei identified at longer wavelengths using a variety of techniques. One of these campaigns, carried out by William Brandt at Pennsylvania State University and his collaborators seeks to study suspected supermassive black holes uncovered by the Sloan Digital Sky Survey, using the high X-ray sensitivity of Chandra. Thus far, they have determined the X-ray properties of nearly 60 such objects in locations where structure first appeared in the universe. One of their most important early conclusions is that, according to the new data, the early supermassive black holes fed and grew in much the same way as those now active closer to Earth.

6.3   THE UNIVERSE AGLOW

And yet, these barely audible X-ray murmurs speak only of those particular supermassive black holes whose orientation facilitates the transmission of their high-energy radiation toward our detectors. Their actual number must be higher than even precision instruments such as the Hubble and Chandra observatories can reveal. Indeed, there is now growing evidence that many – perhaps the majority – of the supermassive black holes in the universe are obscured from view.

Astronomers have puzzled for years over the possible origin of the faint X-ray background pervading the intergalactic medium. Unlike the cosmic microwave background radiation left over from the Big Bang, the photons in the X-ray haze are too energetic to have been produced near the beginning of time. Instead, this radiation field suggests a more recent provenance associated with a population of sources whose overall radiative output may actually dominate over everything else in the cosmos. Stars and ordinary galaxies simply do not radiate profusely at such high energy, and therefore cannot fit the suggested profile.

Quasars would do nicely, but a simple census shows that, in order to produce such an X-ray glow, for every known quasar there ought to be ten more obscured ones. This would also mean that the growth of most supermassive black holes by accretion is hidden from the view of optical, ultraviolet, and near infrared telescopes.

Attempts at uncovering these reticent giants are therefore banking on the possibility that some of their X-ray photons can elude the surrounding muck and escape in numbers sufficient for our new sensitive instruments to detect them. Whereas optical and ultraviolet radiation from the plasma falling into the black hole is absorbed by nearby gas and dust, the higher energy X-rays are only partially attenuated, offering some hope that the hot obscured cauldron at the center of its host galaxy may be seen flickering after all.

Finally, the search has paid off. Investigators from the University of Cambridge, the University of Durham, University College London, and l'Observatoire Midi-Pyrénées in Toulouse have reported the discovery of an object they call a Type-2 quasar. Invisible to optical light telescopes, the nucleus of this otherwise normal looking galaxy betrayed its supermassive guest with a glimmer of X-rays delivered across a 6 billion-light-year chasm to Earth.[2] These scientists suggest that many more quasars, and their supermassive black-hole power sources, may be hidden in otherwise innocuous-looking galaxies. But sensitive X-ray detectors, like those installed on the Chandra observatory, can sense their faint X-ray gleam and expose them for what they really are – typical quasars seen at an inopportune angle.

In an elegant confirmation of this result, investigators from the Harvard-Smithsonian Astrophysical Observatory and the Ohio State University used Chandra to peer through the enshrouding clouds of ten other obscured suspected quasars to reveal the same black-hole signature – hot plasma glowing in X-rays.[3] Quasars, and the blanketed nuclei of many galaxies, are evidently the same phenomenon, only viewed from different angles. The supermassive black holes uncovered thus far must be merely the tip of the proverbial iceberg.

And so, the all-pervasive X-ray haze, in combination with the discovery of gas-obscured quasars, now point to supermassive black holes as the agents behind perhaps *half* of all the universe's radiation produced after the Big Bang. Ordinary stars no longer monopolize the power as they had for decades prior to the advent of space-borne astronomy.

6.4 FUTURE DIRECTIONS

Looking farther afield, the prospects for learning more about the nature of supermassive black holes look very promising indeed – in *both* the near and distant future. Several major undertakings will improve our imaging and spectroscopic capabilities in both the radio and X-ray portions of the spectrum, and LISA (the Laser Interferometer Space Antenna) will open up a whole new window of opportunity for studying the distortions induced on the fabric of spacetime by violent gravitational interactions. As we have already seen (see Chapter 4), LISA's

---

[2] See Fabian *et al.* (2000).
[3] See Green, Aldcroft, Mathur, *et al.* (2001).

expected launch in 2010 will herald a bright new age of space exploration, stretching our frontier well beyond what radiation can let us see. By detecting gravitational waves undulating from distant black-hole sources, astronomers will be able to sense the behavior of massive objects in the presence of unimaginably strong fields, testing general relativity, and possibly even uncovering flaws that hint at new, more comprehensive descriptions and theories of nature.

The windows to be opened by ARISE (Advanced Radio Interferometry between Space and Earth; see Chapter 5) and more elaborate ground-based millimeter arrays will be equally fascinating and conducive to profound change in our communion with nature. Both of these developments – one stretching the baseline of radio interferometry into space, the other creating a worldwide baseline for interferometry at millimeter wavelengths – are geared toward greatly enhancing the resolving power of instruments designed to probe deeper and deeper into the bottomless well of gravity in supermassive objects. Many astrophysicists suspect that an image of the event horizon in a nearby black hole will be feasible within a matter of only years.[4]

Their impressive stature notwithstanding, existing radio telescopes (see Figs. 5.4 and 5.5) are not all usable at the shorter wavelengths because they cannot maintain sufficient structural integrity to provide a pure millimeter or submillimeter signal. So a major problem with conducting worldwide coordinated observations at these wavelengths is simply the paucity of appropriate sites.

The idea for developing a global network of millimeter telescopes, which has come to be known as CMVA – an acronym derived from Coordinated Millimeter VLBI Array – actually goes back to the mid 1990s, when members of the Haystack Observatory in Massachusetts developed plans to create the network for initial observations at 3 millimeters and additional experimental observations at 1.3 millimeters. Since then, the goal of the CMVA has been to

---

[4] See, for example, Falcke, Melia, and Agol (2000), Bromley, Melia, and Liu (2001), and Melia (2003).

continually break new technological ground for later exploitation at progressively shorter wavelengths. Thus far, up to 12 stations around the world have been able to participate in global VLBI sessions at 3 millimeters, organized twice a year through the CMVA. Although unfavorable weather conditions and technical problems at some sites sometimes affect them, these campaigns are generally successful and provide good observations of compact emitting regions, including the galactic center.

At 1 and 2 millimeters, however, the number of telescopes is much smaller than at 3 millimeters, which greatly reduces the coverage. Thankfully, this situation is rapidly changing. For example, the new Heinrich-Hertz telescope on Mount Graham near Tucson recently participated in a VLBI experiment at 1 millimeter for the first time. Even more exciting is the proposed development of the giant radio telescope known as ALMA, which conveys better than any other project the growing enthusiasm from the world's astronomical community. The Atacama Large Millimeter/Submillimeter Array is conceived as a radio telescope composed of 64 transportable 12-meter-diameter antennas distributed over an area 14 kilometers in extent. In the early part of 2001, representatives from Europe, Japan, and North America met in Tokyo to sign a resolution affirming their mutual intent to construct and operate this facility in cooperation with the Republic of Chile, where the telescope is to be located. ALMA will be built on the Andean plateau at 5000 meters altitude near the Atacama Desert, and is considered to be the first truly global project in the history of fundamental science. The telescope is scheduled to be fully operational in 2010.

X-ray astronomy, on the other hand, must be conducted entirely above Earth's soupy atmosphere. The Chandra satellite – the latest NASA innovation – has merely given astronomers a taste of what X-ray images with exceptional spatial resolution can reveal. Scientists and engineers at the Goddard Space Flight Center in Maryland, at Columbia University in New York, and at CALTECH in Pasadena, among others, are conjuring up one of the most ambitious advances in the history of high-energy astronomy. Taking as their cue the lessons learned from the evolution of ground-based optical telescopes, in which many smaller units working in unison are in the end more powerful and easier to build than one single cumbersome device, these investigators are designing and building the Constellation-X Observatory (see Fig. 6.3). Four individual X-ray telescopes working together will have a combined sensitivity 100 times greater than any past or present X-ray mission.

More imaginative still is a NASA mission now under planning that purports to achieve nothing short of actually photographing the event horizon of several nearby supermassive black holes in X-ray light. A duo of powerful new NASA telescopes, with costs estimated in the billions of dollars, are being developed collaboratively by NASA and the University of Colorado at Boulder, and are proposed for flight before 2020. These telescopes are part of the Microarcsecond X-ray Imaging Mission, or MAXIM for short. The main mission would consist of a fleet of 33 spacecraft, each containing a relatively small telescope. But by combining the data gathered by so many separate instruments distributed over an extraordinarily large baseline in space, one may achieve a resolution of the sky about one *million* times better than what is currently attainable. A ground-based optical telescope with this same capability would enable us to read a newspaper on the lunar surface!

To put this achievement in context, note that at a distance of 60 million light-years, the event horizon of the 3-billion-solar-mass black hole in the nucleus of M87 (see Figs. 5.9 and 5.10) projects a diameter of 5 microarcseconds. MAXIM's intended resolution – the angular separation of features that it can identify – is about one microarcsecond, so future X-ray astronomers will be able to see the dark depression shimmering at the center of this giant elliptical galaxy. But with a projected width of over 30 microarcseconds, the easiest dark pit of all to photograph with MAXIM will be that projected by Sagittarius A* at the heart of the Milky Way.

This technology has its own problems to contend with.[5] The wavelength of an X-ray is about 1000 times smaller than that of visible light, making X-ray telescopes very difficult to build. Surface irregularities that are too small to affect visible light can easily scatter X-rays. In addition, to obtain a true focus, X-ray photons must reflect twice from very carefully figured hyperbolic and parabolic surfaces, nested concentrically in very precise formation. Instead, MAXIM will utilize a method similar to VLBI, in which two or more telescopes are coupled in order to synthetically build an aperture equal to the separation of the individual instruments. Instead of precisely focusing X-rays with expensive mirrors onto a detector, the MAXIM team will use readily made flat mirrors to mix the photons, producing an even sharper image, similar to the way sound waves can be combined to either cancel each other out (resulting in silence) or amplify the sound when one crest adds to the other.

The concept calls for the fleet of smaller telescopes to be spaced evenly in orbit around the perimeter of a circle, the diameter of which will vary from 1 to 10 kilometers, and for the whole assembly to be orbiting about the Sun. From there they would collect X-ray beams and funnel them to a larger telescope stationed at the hub, which could then relay the accumulated data back to Earth, several million miles away.

6.5 IS THE UNIVERSE ITSELF A BIG BLACK HOLE?

The field of black-hole research is clearly in a period of renaissance, with wave upon wave of breathtaking discoveries creating headlines on a regular basis, it seems, and with future missions promising to take us to the edge of validity of current physical laws. Supermassive black holes are no longer the oddity of decades past, but rather a necessity in any comprehensive description of structure in the universe.

---

[5] This work has been spearheaded by Webster Cash and his group at the University of Colorado, in collaboration with NASA's Marshall Space Flight Center in Huntsville, Alabama. They announced their design in the 14 September 2000 issue of *Nature*.

Some astronomers are taking this essential role to a rather daring conclusion, wondering, in fact, if we ourselves may be living inside the
biggest black hole of all – the universe itself. Well, this question is
not really well posed, as we shall soon see, but it does make for some
intriguing reflection on cause and effect, and on the origin of all things.

A black hole is a parcel of closed spacetime embedded within a
larger space (and time) that may contain matter, radiation, and probably other black holes as well. On the other hand, the universe as we
know it is all encompassing, so for us to view it as a black hole, it
would be necessary to hypothesize the existence of an undetected –
and probably forever undetectable – hyperspace within which it is
ensconced.

The major difficulty in maintaining a scientific posture with
this discourse is that physicists do not yet have a complete theory
unifying all the fundamental forces of nature at the instant of the Big
Bang. They can say with some precision what transpired 10^{-43} second
later, and any time thereafter, but that first fleeting moment borders
on philosophy and aesthetics, not the rigor of verifiable hardcore science. For example, there is no possibility of linking current theories to
experimentation with the early universe – that is, we cannot simply
"build" another cosmos – so our theorizing must be accepted or re-
jected primarily on the basis of pure reasoning, and perhaps the power
of prediction at later times.

The most unsettling, yet the most engaging, aspect of the Big
Bang is the problem of beginning – the apparent singularity from which
expansion started. An initial state of arbitrarily high density seems to
be inescapable, just as catastrophic gravitational collapse evidently
squeezes to zero volume matter falling into a bottomless well of gravity. In principle, understanding the process of gravitational contraction may resolve the mystery of our distant past, perhaps revealing
new laws of physics along the way.

Still, certain issues pertaining to the question of the universe as
a black hole may already be addressable within the current framework.
Questions such as "Does the universe lie within its own gravitational

radius, i.e., within its own event horizon?" and "What happens toward zero time in the current universe should we reverse the clock?" can at least be broached with the language of scientific principles already recognized and tested.

It may seem surprising to hear that the average density of matter within a black hole need not be extraordinarily large. Its value depends critically on how big the object is. The problem is simply to get enough material within a given radius to produce an event horizon at that radius. From Chapter 3, we recall that the Schwarzschild radius is $2GM/c^2$, so in effect the black hole's size scales directly with its total mass M. But for a given value of M, its density drops off inversely as the enclosed volume, which is proportional to the radius cubed. Thus, ponderous black holes actually have a significantly lower density than their lighter brethren. For example, if a 100-kilogram person were to suddenly shrink to black hole proportions, he would need to have a radius no bigger than about 10^{-23} centimeter, but his density would then rise to the extraordinary value of 10^{73} grams per cubic centimeter. The Sun, squeezed into a black hole, would have a 3-kilometer radius, but its density would be only 10^{16} grams per cubic centimeter.

Now consider what happens as we increase the mass further, to a value not unlike that of a typical supermassive black hole in the nucleus of an active galaxy. For a 100-million-solar-mass object, its Schwarzschild radius grows to 2.4 hundred million kilometers – roughly the size of Mars's orbit about the Sun. But its average density is incredibly only about 1 gram per cubic centimeter – the density of water!

An extremely large region of space, such as the universe, does not have to be very densely filled with matter in order to create curved light paths or even to entomb spacetime itself by forming an event horizon. Given that we see the universe from "inside," how does one then go about determining whether it is above its black-hole density or not? Part of the answer actually goes back to the work of Sir Isaac Newton who, in order to describe the moon's motion around the Earth, used the newly invented calculus to prove a very important theorem for his universal law of gravitation. He showed that the gravitational field outside a spherically symmetric body behaves as if the whole mass were concentrated at its center. In other words, the moon feels exactly the same gravitational influence from the Earth as it would from an object with the same mass, though only the size of an apple situated at the center of where Earth now stands.

In 1923, not long after general relativity was established, George Birkhoff (1884–1944) made the surprising discovery that Newton's theorem was valid even for this more comprehensive description of gravity, though with some appropriate corrections. He demonstrated that even if a spherically symmetric body were collapsing or expanding radially, the Schwarzschild metric describing its gravitational field in empty space would not change in time. In other words, the effect of gravity outside a spherically symmetric body does not depend on how big that object is – it is based solely on how much mass is enclosed within its surface.

The Birkhoff theorem seemed peculiar because in general relativity a nonstatic body generally radiates gravitational waves. We now know that in fact no gravitational radiation can escape into empty space from an object that looks the same from all directions, unlike the pair of black holes orbiting about each other in Fig. 4.5. His result may be applied with equal validity inside an empty spherical cavity at the center of a spherically symmetric (though not necessarily static) body. Here, however, there is no enclosed mass at any point within the cavity so, according to his theorem, there is no gravitational field anywhere inside it.

The value in Birkhoff's work is that, under the assumption of uniformity, we can calculate the gravitational field anywhere in the universe *relative* to another point a distance d away, by simply estimating how much mass is enclosed within a spherically symmetric volume of radius d centered on that other point. For the sake of specificity, let us just put ourselves in the middle and see how far out we need to go before we hit the universe's event horizon.

According to Hubble's discovery of an expanding universe back in the 1920s and 1930s, distant objects are receding from us with a velocity proportional to their distance. It turns out that this rate of recession approaches the speed of light for matter 12 billion light-years away, and this must therefore be the radius of that part of the universe with which we have interacted via influences that travel at the speed of light. (Two specific examples are electromagnetic and gravitational waves.) It is what astronomers call the size of the *visible* universe.

Birkhoff's theorem tells us that the average internal density required to produce an event horizon at 12 billion light-years is about 5×10^{-30} grams per cubic centimeter – an incredibly small number, the equivalent of only six hydrogen atoms per cubic meter. Even so, it exceeds the best current estimates astronomers have made by a factor of roughly three to five, depending on which newspaper vendor you talk to. Could the dark energy invoked to explain the universe's acceleration make up the difference (see Chapter 3)? Without it, the visible universe could not be a black hole in the strictest sense of the term, though it would come alarmingly close. Let us think about this for a moment. Of all the possible average densities that the universe could have had, why is it that the one with which it is apparently endowed is so strikingly close to the value needed to create an event horizon at the edge of what is visible?

Perhaps the answer lies in another important consideration we have so far ignored in this discussion. According to current cosmological models, the expansion of the universe is driven not by matter moving through space, but rather by the stretching of space itself. This is more than just an idle concept since the very idea of inflation depends critically on the validity of an expanding space, and without inflation (see Chapter 3), many problems with the basic Big Bang model would go unsolved. The expansion of space, however, can proceed faster than the speed of light. The postulates of special relativity do not apply to this phenomenon, since they only specify what the maximum speed of transmission *through* the space can be, and that is the speed of light. So although we may not be able to see the "rest" of the cosmos beyond the visible limit at 12 billion light-years, it may nonetheless be there and expanding in concert with our own visible universe.

Can we therefore extend the radius of our Birkhoff sphere and intersect an event horizon by going beyond the "visible" limit? Well, no. For one thing, if this region is beyond the visible edge of the universe, then it is forever inaccessible to us, and we to it. The influence of gravity cannot travel faster than light either, so whatever mass is present there would never have communicated with the universe we can see, and they could never conspire to pool their influence and produce a common event horizon.

Nonetheless, the answer to the question "Is the universe itself a big black hole?" is a qualified "yes" because of several truly amazing observations completed by an international team of astronomers using the BOOMERanG experiment in 2000. We already touched on the significance of their findings in Chapter 4, but let us now revisit this discovery in the context of the present topic.

Designed to study the cosmic microwave background radiation with unprecedented accuracy, BOOMERanG surveyed 2.5 percent of the sky with an angular resolution of 0.25 degrees during a ten-day balloon flight over Antarctica. This microwave telescope was built to measure fluctuations in the background radiation (see Fig. 4.1) driven by pressure variations propagating throughout the nascent universe. A peak in the frequency of these variations was expected to occur 300 000 years after the Big Bang, when the matter and radiation ceased to interact via photon scattering. Earlier calculations had shown that a universe with a current average density of 5×10^{-30} grams per cubic centimeter would have produced fluctuations with a characteristic angular separation of about 0.75 degrees, well within BOOMERanG's resolving capability.

The team of astronomers who conducted this investigation, led by Paolo de Bernardis of the University of Rome and Andrew Lange of CALTECH, reported that BOOMERanG not only confirmed a primordial origin for the fluctuations, but also clearly identified a peak precisely where these predictions had placed it. The location of the peak means that the density of matter in the universe is within a statistically determined error of only 10 percent of its critical value.

Physicists already know that the combined density of visible and dark matter, and radiation, amounts to only about one-third of the required 5×10^{-30} grams per cubic centimeter. So the rest of it must be the "dark energy" inferred from the accelerated expansion of the universe. Although the evidence for this phenomenon is still rather tentative,[6] cosmologists find it very gratifying that together with the completely independent determination rendered by BOOMERanG, they now paint a self-consistent picture. The cosmos is evidently dominated by dark energy, but in such a way that its overall equivalent mass density is precisely 5×10^{-30} grams per cubic centimeter. The universe, it seems, has an event horizon with a radius of 12 billion light-years, right at the edge of what we can see before the velocity of expansion exceeds the speed of light.

This universe, however, has no apparent singularity right now – its mass is spread out everywhere. Could it be that the Big Bang was nothing more than the initial collapse of the universe to something approaching a point, followed by a bounce? Yes, it's possible, but we may never know for sure because the first 10^{-43} second of the expansion is completely unresolvable with current scientific methods. Let us reverse the clock, and see how far back our present knowledge can take us toward the beginning, and why this interval of 10^{-43} second, known as the Planck time, appears to be impenetrable.

The shortest interval of time that can be probed with current physical laws pushes their applicability to the limits set by three so-called fundamental constants of nature. These are the *measured* values of quantities that characterize the strength of gravity, the speed of light, and the fuzziness of quantum mechanics. Physicists assume

---

[6] See Brian P. Schmidt *et al.* (1998) and Saul Perlmutter *et al.* (1999).

that these quantities are constants in time, in the absence of any evidence to the contrary.

Quantum mechanics argues that we can never be entirely sure of a particle's position or its energy, because in order for us to even know of its existence we must disturb it to sense its presence. Thus, there should always be some positional uncertainty, or an imprecision in energy and time, and any description of the particle's physical behavior must therefore acquire some minimal level of "fuzziness." In our everyday lives, we develop the illusion of precision only because the fuzziness induced by these uncertainties is very small, and our mind clings to the apparent clarity of the outside world as a convenient simplification of the way things really are. Certainly, on a macroscopic scale, this fuzziness does not manifest itself readily, and our description of nature using exact positions and times is quite adequate for our need to interpret much of the activity in our environment. But on a microscopic scale, this fuzziness is paramount, and nothing can happen without the consequences of the implied imprecision.

The uncertainty in the particle's position is characterized by Planck's constant, h. The Planck length – the shortest distance we can probe – depends on how strong the effect of gravity is on such scales. This in turn is specified by the gravitational constant, G, in Newton's universal law of gravitation. The bigger this coupling constant is, the stronger is the attraction between two given masses. The Planck time is then the interval of time required to communicate information across this distance, given that the apparent maximum rate of transmission is the speed of light, c. Together, these constants yield the shortest physical time, $(Gh/c^5)^{\frac{1}{2}}$ (which is approximately 10^{-43} second), that anyone (or anything) can sample.

However, cosmologists do have some confidence in beginning to describe the expansion of the universe from 10^{-43} second onwards. This is where our quantum physics has meaning, because on this level the Schwarzschild radius from general relativity first becomes equal to the smallest scale permitted by the quantum fuzziness, roughly

10^{-33} centimeter, which is still much smaller than the nucleus of an atom. But there is still some remaining uncertainty because physicists diverge in their views of how one should best describe the universe at this point. They still do not know if extra dimensions exist (see Chapter 3), or if string theory is correct. One view has it that during the Planck era (when the universe was about 10^{-43} second old), the cosmos should best be described as a quantum "foam" of ten spatial dimensions containing Planck-length-size black holes, continuously being created and annihilated, with no cause or effect. The reason for the latter is that, on quantum scales, particles can be created without the conservation of energy, as long as they exist only fleetingly so that the violation falls within the uncertainty prescribed by Planck's constant.

One of the reasons our physics is incomplete near the Planck era is related to the hierarchy problem we discussed in Chapter 3. Science does not yet provide a description of how the forces of nature unify during this time. At the excruciatingly high energies and temperatures prevalent then, the forces of nature would have become symmetric, meaning that they would have resembled each other and would have acquired a similar strength – they would have *unified* into a single entity. Physicists are actively pursuing the grail of grand unification of all four forces, and have already achieved some notable success in this pursuit. Toward the end of the twentieth century, the interactions due to the weak and electromagnetic forces were framed into a single phenomenon known as the electroweak force by Sheldon Glashow, Steven Weinberg, and Abdus Salam, who were awarded the Nobel Prize in Physics for this effort in 1979.

The weak force, which is mediated by very heavy particles known as W and Z, is responsible for the transformation of a neutron into a proton within the nucleus of an atom, whereas the electromagnetic force provides an interaction between charged particles, such as the electron and a proton. At the time of their discovery in 1983, the W and Z particles were the most massive known – each weighing in at almost 90 times the mass of the proton – whereas the photon, the carrier of the electromagnetic force, is massless. The unification of these two forces occurs when the energy available for the process is so high that even this enormous mass difference between the two sets of carriers becomes inconsequential. In the early universe, this would have been the situation until the ambient temperature dropped below about 10^{15} Kelvin, after which the mass difference would have split the rates at which these particles could interact, thereby creating the appearance of two independent forces.

Attempts are now underway to unify the strong and electroweak forces, a process known as Grand Unification, but this is proving to be much more challenging, in part because what is required is the conversion of certain particles, such as electrons, into completely different types of entities, known as quarks. This unification, if possible, would result in a split of the rates of interaction when the temperature in the early universe dropped below about 10^{27} Kelvin, much closer to the Planck era.

The final unification, between the electroweak, strong and gravitational forces, is well beyond the realm of study with earthbound experiments, because the energies and temperatures required to approach the necessary scale of interaction are simply unreachable. It may seem peculiar, but learning more about the early universe may actually be necessary for this branch of particle physics to make progress of its own toward a "complete" understanding of what governs the substance and behavior of particles.

These unknowns impact the cosmologists' view progressively more and more, as they labor closer and closer to the Planck scale. The exploration terminates – indefinitely it would seem – at 10^{-43} second. Only the development of a completely new, overarching description of nature that obviates the fuzziness of quantum mechanics could change this situation. Still, physicists are a clever lot, so there is always hope. Is the Universe itself a big black hole? It now seems that the answer is yes, but how and why it got that way persist as the most profound mysteries in nature.

6.6 ULTIMATE FATE

Counterposing the uncertainty of what transpired at the very beginning of the Big Bang, the question of how the universe will play itself out may be easier to address, though, as always, the story unfolds through the prism of human perception and interpretation. It would be utterly presumptuous and self-debilitating for us to view this prognostication as absolute and fully written. On the contrary, it is an evolving narrative, likely to be swayed by many future developments and discoveries in particle physics and astronomy.

For now, the three leading characters in this play are the total mass enclosed within the visible universe, the Grand Unified Theory (GUT) that will ultimately account for the unification of all known forces, and Hawking radiation. Up until the era when the reservoir of primordial matter – primarily hydrogen and other light elements – is fully exhausted, stars will continue to form and galaxies will collide and grow. Looking into the future, however, matter will ultimately partition itself into several quasi-terminal states, among them dying stellar embers, white dwarfs and neutron stars, asteroids and planets, and tenuous gas dispersed throughout the cosmos. But regardless of what the eventual configuration will be, life as we know it will not be viable forever. Without the energy released from nuclear burning, life-sustaining environments will become untenable. In the meantime, supermassive black holes will continue to grow as clump after clump of gas succumbs to the inexorable inward pull of gravity, adding to the total mass entombed below the growing number of event horizons.

Life will undoubtedly evolve considerably and survive much longer than we could now imagine. In the absence of nucleosynthesis, our descendants may even find a way of using energy liberated by accretion onto black holes in order to power their survival. But certain processes predicted by the GUT will change the universe dramatically and irreparably, making any such attempts futile in the long run. In these theories, all sorts of particles can (and must) mutate into other entities, a process that may be induced by either collisions or spontaneous self-decay. A proton, for example, will eventually split into a positron (the electron's antiparticle) and a pion, the particle that helps to mediate the nuclear force. Neutrons are already known to be unstable; in a matter of only minutes, they decay into protons (this process is induced by the weak force), so they too must eventually split into sub-components. By permitting this conversion – nay, requiring it – the GUT will guarantee that the two most significant constituents of atomic nuclei will be removed permanently from the composition table. Diamonds are not forever!

Physicists still do not know the mass of several particles that mediate the unified force, so the time required for protons and neutrons to decay is uncertain. The best current estimates endow the proton with an expected lifetime somewhere between 10^{32} and 10^{41} years.[7]

By this time, galaxy collisions (see Chapter 4) will have been relegated to ancient history by the expansion of the universe, which would have continued to drive the participants apart. Supermassive black holes will therefore stop growing some day because they will have absorbed all the limited supply of matter in their environment. Estimates place the terminal mass of these objects somewhere between 1 billion and 10 billion Suns.

The universe in this era will be completely unrecognizable to sentient beings living now, since it would have mutated to the point where life itself would be impossible. As best as physicists can tell, the cosmos will be an extremely thin dark veil of fundamental particles, such as electrons, positrons, neutrinos, and highly redshifted photons. Very few atoms will be left, and these too will eventually vanish as their constituent protons and neutrons disintegrate. And floating aimlessly through this enormous sea of virtually nothing will be the ensemble of billion-solar-mass black holes roaming freely for a near eternity, sucking up whatever scant morsels they encounter.

---

[7] A full discussion of the relevant parameters and other considerations may be found in Adams and Laughlin (1997).

Evidently, supermassive black holes appeared early in the history of the universe and will stay late – very late. After 10^{32} to 10^{41} years, they will be the *only* structures of any significance left in the cosmos. But in what appears to be the final act of fair play, even they will not exist forever. Once black holes stop growing, they slowly begin to shrivel via a loophole created by the application of quantum mechanics, a theory that is known to be correct, if not complete. General relativity is a classical theory, operating on the basis of precise measurements of physical quantities, such as distance and time. The very notion of defining an event horizon makes sense as long as we can precisely place this surface and particles around it at perfectly known locations. But quantum mechanical fuzziness requires some positional uncertainty, or an imprecision in energy and time. Physicists are therefore uncomfortable with the idea of a perfectly localized and sealed event horizon, since these notions completely ignore the quantum mechanical uncertainty on the smallest scales.

A phenomenon discovered in 1974 by Stephen Hawking may be the first step in the eventual resolution of this problem.[8] The name itself, *quantum* mechanics, reveals the essence of the physical description on a microscopic scale. It tells us that at this level all measurable entities are to be thought of as comprising tiny bundles (or quanta) of "something," which in the case of light are known as photons. In the appropriate terminology, one says that fluctuations in a field, say the gravitational field, are associated with the manifestation of these quanta, which can appear or vanish as the fluctuations grow or subside. The connection between these bundles and the fuzziness is that their size, energy, and lifetime are directly related to the scale of the imprecision, that is, how fuzzy the measurements of position or energy turn out to be.

Quanta such as photons bubble up spontaneously out of vacuum if an adequate source of energy lies nearby. But a crucial fact that we

---

[8] Readers who would like to learn more about the technical aspects of this phenomenon, and the evaporation of black holes in general, will find the discussion in Thorne, Price, and Macdonald (1986) very helpful. See also Wald (1984).

have gathered from the observed behavior of these fields is that when the bundles materialize spontaneously, they always do so in pairs, as if something must be split in order to create the fluctuation. So a quantum, or particle, with negative charge can only materialize if at the same time its counterpart, with positive charge, also comes into being. Given that every characteristic we can assign to this bundle must be matched by the opposite attributes of its partner particle, it makes sense then to talk of these as particles and antiparticles, or matter and antimatter.

The phenomenon discovered by Hawking[9] is directly associated with this creation of quantum particles in vacuum due to fluctuations in the gravitational field of the black hole. Particles created in this way live fleetingly and then annihilate with each other's counterpart to re-establish the vacuum after the fluctuation has subsided. We note, however, that fluctuations in the gravitational field of the black hole have a wavelength commensurate with its size. So when these fluctuations manifest themselves as photons, or any other type of particle whose rest mass is small compared to the amplitude of the fluctuation, their wavelength, too, corresponds to the size of the black hole. The fleeting quanta produced beyond the event horizon of very massive black holes are therefore much redder, and hence of lower energy, than those associated with their smaller brethren.

The paired quanta produced in this fashion annihilate outside the event horizon very quickly (in about one-millionth of a millionth of a millionth of 1 second). But some pairs, argued Hawking, will have a member that dips below the membrane of no return, abandoning its partner to the whim of the outside universe. Without a partner to annihilate, the detached particle flees the black hole's sphere of influence and merges into the flux of escaping radiation headed for infinity. To an observer on Earth, this looks like the black hole is actually radiating, though the mechanism is clearly indirect. Nevertheless, the

---

[9] Some of Hawking's early discussion on this topic appeared in a paper published by *Nature* in 1974.

source of energy for these fleeing particles is ultimately the black hole itself, and although we cannot claim that the radiation originated from within the event horizon, its energy surely did, and the dark object pays the price with a consequent decrease in its mass. If this simple application of quantum mechanics survives the test of time, it appears that all black holes must evaporate eventually.

The Hawking radiation from a black hole with barely the mass of 30 Suns has such a long wavelength, and is therefore so feeble, that it would take such an object 10^{61} times the current age of the universe to evaporate completely. But after 10^{98} years, even the 100-billion-solar-mass behemoths will be gone, completely and forever – the final act of fair play. And thus will end the saga of the most powerful objects in the universe, facing eternity as ghosts in a lifeless darkness.

References

Adams, F. C., and G. Laughlin (1997), "A Dying Universe: The Long Term Fate and Evolution of Astrophysical Objects," *Reviews of Modern Physics* 69, 337–372.

Allen, C. W. (1991), *Astrophysical Quantities*, Athlone Press.

Alpher, R., and R. Herman (1948), "Evolution of the Universe," *Nature* 162, 774–775.

Anderson, J. L. (1967), *Principles of Relativity Physics*, Academic Press.

Arkani-Hamed, N., S. Dimopoulos, and G. Dvali (1998), "The Hierarchy Problem and New Dimensions at a Millimeter," *Physics Letters B* 429, 263–272.

Baade, W., and R. Minkowski (1954), "On the Identification of Radio Sources," *Astrophysical Journal* 119, 215.

Backer, D. C., and R. A. Sramek (1999), "Proper Motion of the Compact, Non-thermal Radio Source in the Galactic Center, Sagittarius A*," *Astrophysical Journal* 524, 805–815.

Balberg, S., and S. L. Shapiro (2002), "Gravothermal Collapse of Self-Interacting Dark Matter Halos and the Origin of Massive Black Holes," *Physical Review Letters* 88, 101301.1–101301.4.

Balick, B., and R. L. Brown (1974), "Intense Sub-arcsecond Structure in the Galactic Center," *Astrophysical Journal* 194, 265–270.

Bernstein, H. J., and A. V. Pillips (1981), "Fiber Bundles and Quantum Theory," *Scientific American* 245, 123–137.

Binney, J., and S. Tremaine (1987), *Galactic Dynamics*, Princeton University Press.

Birkhoff, G. (1923), *Relativity and Modern Physics*, Harvard University Press.

Blandford, R. D., C. F. McKee, and M. J. Rees (1977), "Superluminal Expansion in Extragalactic Radio Sources," *Nature* 267, 211–216.

Blandford, R. D., and M. J. Rees (1978), "Extended and Compact Extragalactic Radio Sources. Interpretation and Theory," *Physica Scripta* 17, 265–274.

Blandford, R. D., and R. L. Znajek (1977), "Electromagnetic Extraction of Energy from Kerr Black Holes," *Monthly Notices of the Royal Astronomical Society* 179, 433–456.

Bondi, H. (1980), *Relativity and Common Sense*, Dover.

Bower, G. A., D. O. Richstone, G. D. Bothun, and T. M. Heckman (1993), "A Search for Dead Quasars Among Nearby Luminous Galaxies. I. The Stellar Kinematics in the Nuclei NGC 2613, NGC 4699, NGC 5746, and NGC 7331," *Astrophysical Journal* 402, 76–94.

Bromley, B., F. Melia, and S. Liu (2001), "Polarimetric Imaging of the Massive Black Hole at the Galactic Center," *Astrophysical Journal Letters* 555, L83–L87.

Burbidge, G. R., T. W. Jones, and S. L. Odell (1974), "Physics of Compact Nonthermal Sources III. Energetic Considerations," *Astrophysical Journal* 193, 43–54.

Carilli, C. L., and P. D. Barthel (1966), "Cygnus A," *The Astronomy and Astrophysics Review* 7, 1–54.

Chandrasekhar, S. (1931), "The Maximum Mass of Ideal White Dwarfs," *Astrophysical Journal* 74, 81–82.

Colbert, E. J. M., and A. F. Ptak (2002), "A Catalog of Candidate Intermediate-Luminosity X-Ray Objects," *Astrophysical Journal Supplements* 143, 25–45.

Cunningham, C. T., and J. M. Bardeen (1973), "The Optical Appearance of a Star Orbiting an Extreme Kerr Black Hole," *Astrophysical Journal* 183, 237–264.

Damour, T., in *Three Hundred Years of Gravitation*, S. W. Hawking and W. Israel (eds.) (1987), Cambridge University Press, 128–198.

Davydov, A. S. (1976), *Quantum Mechanics*, Pergamon Press.

de Bernardis, P., P. A. R. Ade, J. J. Bock, *et al.* (2000), "A Flat Universe from High-Resolution Maps of the Cosmic Microwave Background Radiation," *Nature* 404, 955–959.

Dicke, R. H., P. J. E. Peebles, P. G. Roll, and D. T. Wilkinson (1965), "Cosmic Black-Body Radiation," *Astrophysical Journal* 142, 414–419.

Drake, S. (1981), *Cause, Experiment, and Science*, Chicago University Press.

Duschl, W. J., and H. Lesch (1994), "The Spectrum of Sagittarius A* and its Variability," *Astronomy and Astrophysics* 286, 431–436.

Eddington, A. S. (1935), "Minutes of a Meeting of the Royal Astronomical Society," *Observatory* 58, 37.

Eisenstein, D. J., and A. Loeb (1995), "Origin of Quasar Progenitors from the Collapse of Low-spin Cosmological Perturbations," *Astrophysical Journal* 443, 11–17.

Fabian, A. C., I. Smail, K. Iwasawa, *et al.* (2000), "Testing the Connection Between the X-ray and Submillimeter Source Populations Using Chandra," *Monthly Notices of the Royal Astronomical Society* 315, L8–L12.

Falcke, H., F. Melia, and E. Agol (2000), "Viewing the Shadow of the Black Hole at the Galactic Center," *Astrophysical Journal Letters* 528, L13–L17.

Fermi, E. (1956), *Thermodynamics*, Dover.

Ferrarese, L., and D. Merritt (2000), "A Fundamental Relation Between Supermassive Black Holes and Their Host Galaxies," *Astrophysical Journal Letters* 539, L9–L12.

Feynman, R. P., and S. Weinberg (1989), *Elementary Particles and the Laws of Physics*, Cambridge University Press.

Field, J. V. (1987), *Kepler's Geometrical Cosmology*, University of Chicago Press.

Fromerth, M. J., and F. Melia (2001), "The Formation of Broad-Line Clouds in the Accretion Shocks of Active Galactic Nuclei," *The Astrophysical Journal* 549, 205–214.

    (2000), "Determining the Central Mass in Active Galactic Nuclei Using Cross-Correlation Lags and Velocity Dispersions," *The Astrophysical Journal* 533, 172–175.

Gamow, G. (1948), "Evolution of the Universe," *Nature* 162, 680–682.

Garrett, M., T. W. B. Muxlow, S. T. Garrington, *et al.* (2001), "AGN and Starbursts at High Redshift: High-Resolution EVN Radio Observations of the Hubble Deep Field," *Astronomy and Astrophysics* 366, L5–L8.

Gebhardt, K., J. Kormendy, L. C. Ho, *et al.* (2000), "A Relationship Between Nuclear Black Hole Mass and Galaxy Velocity Dispersion," *Astrophysical Journal Letters* 539, L13–L16.

Gebhardt, K., R. M. Rich, and L. C. Ho (2002), "A 20,000 Solar Mass Black Hole in the Stellar Cluster G1," *Astrophysical Journal Letters* 578, L41–L45.

Giacconi, R., A. Zirm, J.-X. Wangs, *et al.* (2002), "Chandra Deep Field South: The 1 Ms Catalog," *The Astrophysical Journal Supplements* 139, 369–410.

Gillispie, C. C. (1997), *Pierre-Simon Laplace, 1749–1827: A Life in Exact Science*, Princeton University Press.

Green, P. J., T. L. Aldcroft, S. Mathur, B. J. Wilkes, and M. Elvis (2001), "A Chandra Survey of Broad Absorption Line Quasars," *The Astrophysical Journal* 558, 109–118.

Greene, B. (2000), *The Elegant Universe*, Vintage Books.

Gundlach, J. H., and S. M. Merkowitz (2000), "Measurement of Newton's Constant Using a Torsion Balance with Angular Acceleration Feedback," *Physical Review Letters* 85, 2869–2872.

Guth, A. (1981), "Inflationary Universe: A Possible Solution to the Horizon and Flatness Problems," *Physical Review D* 23, 347–358.

Guthrie, W. K. C. (1962), *A History of Greek Philosophy, Volume I: The Earlier PreSocratics and the Pythagoreans*, Cambridge University Press.

Gwinn, C. R., R. M. Danen, T. Kh. Tran, J. Middleditch, and L. M. Ozernoy (1991), "The Galactic Center Radio Source Shines Below the Compton Limit," *Astrophysical Journal Letters* 381, L43–L46.

Haehnelt, M. G., and M. J. Rees (1993), "The Formation of Nuclei in Newly Formed Galaxies and the Evolution of the Quasar Population," *Monthly Notices of the Royal Astronomical Society* 263, 168–178.

Hanany, S., P. Ade, A. Balbi, *et al.* (2000), "MAXIMA-1: A Measurement of the Cosmic Microwave Background Anisotropy on Small Angular Scales," *Astrophysical Journal Letters* 545, L5–L8.

Hawking, S. W. (1974), "Black Hole Explosions?" *Nature* 248, 30.

Hollywood, J. M., and F. Melia (1995), "The Effects of Redshifts and Focusing on the Spectrum of an Accretion Disk in the Galactic Center Black Hole Candidate Sgr A*," *Astrophysical Journal Letters* 443, L17–L21.

   (1997), "General Relativistic Effects on the Infrared Spectrum of Thin Accretion Disks in AGNs; Application to Sgr A*," *Astrophysical Journal Supplements* 112, 423–455.

Hollywood, J. M., F. Melia, L. M. Close, D. W. McCarthy, Jr., and T. A. Dekeyser (1995), "General Relativistic Flux Modulations in the Galactic Center Black Hole Candidate Sgr A*," *Astrophysical Journal Letters* 448, L21–L25.

Hoyle, F., and G. R. Burbidge (1966), "On the Nature of the Quasi-Stellar Objects," *Astrophysical Journal* 144, 534.

Junor, W., J. A. Biretta, and M. Livio (1999), "Formation of the Radio Jet in M87 at 100 Schwarzschild Radii from the Central Black Hole," *Nature* 401, 891–892.

Kerr, R. P. (1963), "Gravitational Field of a Spinning Mass as an Example of Algebraically Special Metrics," *Physical Review Letters* 11, 237–238.

Khokhlov, A., and F. Melia (1996), "Powerful Ejection of Matter from Tidally Disrupted Stars Near Massive Black Holes and a Possible Application to Sgr A East," *Astrophysical Journal Letters* 457, L61–L64.

Kormendy, J., and D. Richstone (1995), "Inward Bound – The Search for Supermassive Black Holes in Galactic Nuclei," *Annual Reviews of Astronomy and Astrophysics* 33, 581–624.

Landau, L. D. (1932), "On the Theory of Stars," *Physikalische Zeitschrift der Sowjetunion* 1, 285–288.

Lee, H. C. (1984), *An Introduction to Kaluza-Klein Theories*, World Scientific.

Linde, A. (1990), *Particle Physics and Inflationary Cosmology*, Harwood Academic Publishers.

Liu, S., and F. Melia (2001), "New Constraints on the Nature of Radio Emission in Sagittarius A*," *The Astrophysical Journal Letters* 561, L77–L80.

   (2002), "An Accretion Induced X-Ray Flare in Sagittarius A*," *The Astrophysical Journal Letters* 566, L77–L80.

Longair, M. S., M. Ryle, and P. A. G. Scheuer (1973), "Models of Extended Radio Sources," *Monthly Notices of the Royal Astronomical Society* 164, 243.

Lynden-Bell, D. (1969), "Galactic Nuclei as Collapsed Old Quasars," *Nature* 223, 690.

Lynden-Bell, D., and M. J. Rees (1971), "On Quasars, Dust and the Galactic Centre," *Monthly Notices of the Royal Astronomical Society* 152, 461.

Lynden-Bell, D., and R. Wood (1968), "The Gravothermal Catastrophe in Isothermal Spheres and the Onset of Red-giant Structure for Stellar Systems," *Monthly Notices of the Royal Astronomical Society* 138, 495.

Mach, E. (1893), *The Science of Mechanics*, trans. by T. J. McCormack, 2nd edn, Open Court Publishing Co.

Malizia, A., L. Bassani, A. J. Dean, *et al.* (2001), "Hard X-Ray Detection of the High-Redshift Quasar 4C 71.07," *The Astrophysical Journal* 531, 642–646.

Martini, P., and R. W. Pogge (1999), "Hubble Space Telescope Observations of the CFA Seyfert 2 Galaxies: The Fueling of Active Galactic Nuclei," *Astronomical Journal* 118, 2646–2657.

Matsumoto, H., *et al.* (2001), "Discovery of a Luminous, Variable, Off-Center Source in the Nucleus of M82 with the Chandra High-Resolution Camera," *Astrophysical Journal Letters* 547, L25–L28.

Maxwell, J. C. (1954), *Treatise on Electricity and Magnetism*, Vol. II, Dover Publications.

McKellar, A. (1941), "The Problems of Possible Molecular Identification for Interstellar Lines," *Publications of the Astronomical Society of the Pacific* 53, 233–235.

Melia, F. (1992), "An Accreting Black Hole Model for Sagittarius A*," *The Astrophysical Journal Letters* 387, L25–L29.

(2001), *Electrodynamics*, The University of Chicago Press.

(2001), "X-ray from the Edge of Infinity," *Nature* 413, 25–26.

(2003), *The Black Hole at the Center of our Galaxy*, Princeton University Press.

Melia, F., B. C. Bromley, S. Liu, and C. K. Walker (2001), "Measuring the Black Hole Spin in Sagittarius A*," *The Astrophysical Journal Letters* 554, L37–L40.

Melia, F., and H. Falcke (2001), "The Supermassive Black Hole at the Galactic Center," *Annual Reviews of Astronomy and Astrophysics* 39, 309–352.

Merrifield, M. R., D. A. Forbes, and A. I. Terlevich (2000), "The Black Hole Mass-galaxy Age Relation," *Monthly Notices of the Royal Astronomical Society* 313, L29–L32.

Merritt, D., and L. Ferrarese (2001), "Black Hole Demographics from the M-Sigma Relation," *Monthly Notices of the Royal Astronomical Society* 320, L30–L34.

Michelson, A. A., and E. W. Morley (1887), "On the Relative Motion of the Earth and the Luminiferous Ether," *American Journal of Science* 34, 333–345.

Mihalas, D., and B. Weibel-Mihalas (1984), *Foundations of Radiation Hydrodynamics*, Oxford University Press.

Miller, A. D., *et al.* (1999), "A Measurement of the Angular Power Spectrum of the CMB from $l = 100$ to 400," *Astrophysical Journal Letters* 524, L1–L4.

Misner, C. W., K. S. Thorne, and J. A. Wheeler (1973), *Gravitation*, W. H. Freeman and Co.

Miyoshi, M., J. Moran, J. Herrnstein, *et al.* (1995), "Evidence for a Black Hole from High Rotation Velocities in a Sub-Parsec Region of NGC 4258," *Nature* 373, 127.

Moore, P. (1996), *The Planet Neptune, An Historical Survey Before Voyager*, Praxis-Wiley.

Morrison, P. "Are Quasi-Stellar Radio Sources Giant Pulsars?" *Astrophysical Journal* 157, L73.

Mushotzky, R. F., L. L. Cowie, A. J. Barger, and K. A. Arnaud (2000), "Resolving the Extragalactic Hard X-ray Background," *Nature* 404, 459–464.

Newton, I. (1966), *Philosophiae Naturalis Principia Mathematica*, trans. by Andrew Motte, revised and annotated by F. Cajori, University of California Press.

Oppenheimer, J. R., and H. Snyder (1939), "On Continued Gravitational Contraction," *Physical Review* 56, 455–459.

Osterbrock, D. E. (2001), *Walter Baade: A Life in Astrophysics*, Princeton University Press.

Ostriker, J. P., and P. J. Steinhardt (2001), "Brave New Cosmos: The Quintessential Universe," *Scientific American* 284, 47–53.

Parker, E. N. (1979), *Cosmical Magnetic Fields*, Oxford University Press.

Penrose, R. (1969), "Gravitational Collapse, the Role of General Relativity," *Rivista Nuovo Cimento* 1 Numero Speciale, 252.

Penzias, A. A., and R. W. Wilson (1965), "A Measurement of Excess Antenna Temperature at 4080 Mc/s," *Astrophysical Journal* 142, 419–421.

Perlmutter, S., G. Aldering, G. Goldhaber, *et al.* (1999), "Measurements of Omega and Lambda from 42 High-Redshift Supernovae," *The Astrophysical Journal* 517, 565–586.

Pound, R. V., and G. A. Rebka (1960), "The Apparent Weight of Photons," *Physical Review Letters* 4, 337–341.

Quinlan, G. D., and S. L. Shapiro (1990), "The Dynamical Evolution of Dense Star Clusters in Galactic Nuclei," *Astrophysical Journal* 356, 483–500.

Rees, M. J. (1966), "Appearance of Relativistically Expanding Radio Sources," *Nature* 211, 468–470.

(1971), "New Interpretation of Extragalactic Radio Sources," *Nature* 229, 312 and 510.

(1988), "Tidal Disruption of Stars by Black Holes of 10^6 to 10^8 Solar Masses in Nearby Galaxies," *Nature* 333, 523–528.

Reid, M. J., A. C. S. Readhead, R. C. Vermeulen, and R. N. Treuhaft (1999), "The Proper Motion of Sagittarius A*. I. First VLBA Results," *Astro-physical Journal* 524, 816–823.

Rybicki, G. B., and A. P. Lightman (1979), *Radiative Processes in Astrophysics*, John Wiley and Sons.

Sakurai, J. J. (1964), *Invariance Principles and Elementary Particles*, Princeton University Press.

Salpeter, E. E. (1964), "Accretion of Interstellar Matter by Massive Objects," *Astrophysical Journal* 140, 796–800.

Scheuer, P. A. G. (1974), "Models of Extragalactic Radio Sources with a Continuous Energy Supply from a Central Object," *Monthly Notices of the Royal Astronomical Society* 166, 513–528.

Schmidt, B. P., N. B. Suntzeff, M. M. Phillips, *et al.* (1998), "The High-Z Supernova Search: Measuring Cosmic Deceleration and Global Curvature of the Universe Using Type IA Supernovae," *The Astrophysical Journal* 507, 46–63.

Schmidt, M. (1963), "3C 273: A Star-like Object with Large Red-shift," *Nature* 197, 1040.

Schödel, R., T. Ott, R. Genzel, *et al.* (2002), "A Star in a 15.2-year Orbit Around the Supermassive Black Hole at the Centre of the Milky Way," *Nature* 419, 694–696.

Shai, A., M. Livio, and T. Piran (2000), "Tidal Disruption of a Solar-Type Star by a Supermassive Black Hole," *Astrophysical Journal* 545, 772–780.

Shapiro, S. L., and S. A. Teukolsky (1992), "Black Holes, Star Clusters, and Naked Singularities: Numerical Solution of Einstein's Equations," *Philosophical Transactions: Physical Sciences and Engineering* 340, 365–390.

Silk, J., and M. J. Rees (1998), "Quasars and Galaxy Formation," *Astronomy and Astrophysics* 331, L1–L4.

Smoot, G. F., and K. Davidson (1993), *Wrinkles in Time*, William Morrow.

Starobinsky, A. A. (1980), "A New Type of Isotropic Cosmological Models without Singularity," *Physics Letters B* 91B, 99–104.

Steinberg, J. L., and J. Lequeux (1963), *Radio Astronomy*, McGraw-Hill.

Strohmayer, T. E. (2001), "Discovery of a 450 Hz QPO from the Microquasar GRO J1655-40 with RXTE," *Astrophysical Journal Letters* 552, 49–53.

Stubbs, P. (1971), "Red Shift Without Reason," *New Scientist* 50, 254–255.

Thorne, K. S. (1995), *Black Holes and Time Warps: Einstein's Outrageous Legacy*, Norton and Co.

Thorne, K. S., R. H. Price, and D. A. Macdonald (1986), *Black Holes: The Membrane Paradigm*, Yale University Press.

Tolman, R. C. (1987), *Relativity, Thermodynamics, and Cosmology*, Dover Publications.

Umemura, M., A. Loeb, and E. L. Turner (1993), "Early Cosmic Formation of Massive Black Holes," *Astrophysical Journal* 419, 459.

van der Marel, R. P., "Structure of the Globular Cluster M15 and Constraints on a Massive Central Black Hole," in *Black Holes in Binaries and Galactic Nuclei*, L. Kaper, E. P. J. van den Heuvel, and P. A. Woudt (eds.) (1999), 246, Springer.

Wald, R. (1984), *General Relativity*, University of Chicago Press.

Wardle, J. F. C., D. C. Homan, R. Ojha, and D. H. Roberts (1998), "Electron-Positron Jets Associated with the Quasar 3C 279," *Nature* 395, 457–461.

Weber, J. (1969), "Evidence for Discovery of Gravitational Radiation," *Physical Review Letters* 22, 1320–1324.

Weinberg, S. (1972), *Gravitation and Cosmology: Principles and Applications of the General Theory of Relativity*, John Wiley and Sons.

(1977), *The First Three Minutes*, Basic Books.

Westfall, R. (1981), *Never at Rest*, Cambridge University Press.

Wheeler, J. A. (1999), *Journey into Gravity and Spacetime*, Freeman and Co.

Williams, R. E., B. Blacker, M. Dickinson, *et al.* (1996), "The Hubble Deep Field: Observations, Data Reduction, and Galaxy Photometry," *Astronomical Journal* 112, 1335.

Wilms, J., C. S. Reynolds, M. C. Begelman, *et al.* (2001), "XMM-EPIC Observation of MCG-6-30-15: Direct Evidence for the Extraction of Energy from a Spinning Black Hole?" *Monthly Notices of the Royal Astronomical Society* 328, L27–L31.

Zel'dovich, Ya. B., and I. D. Novikov (1967), "The Hypothesis of Cores Retarded During Expansion and the Hot Cosmological Model," *Soviet Astronomy* 10, 602.

Zel'dovich, Ya. B., and M. A. Podurets (1965), "The Evolution of a System of Gravitationally Interacting Point Masses," *Astronomicheskii Zhurnal* 42, 963.

Zensus, J. A., and T. J. Pearson (1987), *Superluminal Radio Sources*, Cambridge University Press.

Index